Salisbury Plain

Home of Britain's Military Training

King George VI and Queen Elizabeth visit Larkhill on 3 April 1941. The roar of the guns causes the Queen to cover her ears.

Salisbury Plain

Home of Britain's Military Training

Henry Buckton

Phillimore

First published 2008, this edition 2015

Phillimore & Co. Ltd, an imprint of The History Press
The Mill, Brimscombe Port
Stroud, Gloucestershire, GL5 2QG
www.thehistorypress.co.uk

British Library Cataloguing in Publication Data.
A catalogue record for this book is available from the British Library.

ISBN 978 0 7509 6739 6

Typesetting and origination by The History Press

Contents

List of Illustrations

Acknowledgements and Sources of Reference

Thanks to Rosemary Meeke, editor of *Drumbeat* (the magazine for Tidworth, Netheravon and Bulford Garrison), for permitting the use of articles on the *Drumbeat* website as sources of reference. These include 'Salisbury Plain, An Historical Introduction' by Major E. Clarke-Smith, MA (Cantab), RAEC, written in June 1969, 'A short history of Airfield Camp, Netheravon' and 'The Garrison Theatre'. Thanks to Michael Marshman, County Local Studies Librarian, from the Library and Heritage HQ, Trowbridge, for permitting use of the Wiltshire County Council's Community History website as a source of reference: www.wiltshire.gov.uk/community. Thanks to John Chandler of Hobnob Press for permitting use of the following Hobnob Press titles as sources of reference: *Plain Soldiering. A history of the Armed Forces on Salisbury Plain* by N.D.G. James, copyright N.D.G. James 1987, and *Little Imber on the Down* by Rex Sawyer, copyright Rex Sawyer 2003. Thanks also to Rex Sawyer himself for further information and help. Thanks for permission to quote from *The time has come: Officer and Temporary Gentleman*, by Dennis Wheatley, first published in 1978, kindly provided by Wheatley Productions Ltd, copyright Dennis Wheatley. Thanks to Vic Damon for permission to use the 3rd Armored Division History Foundation website as a source of reference: www.3ad.com, including quotes and photographs. Thanks to Susan Lindsay, curator of the Museum of Army Flying Middle Wallop, for permitting me to use photographs and information contained in the museum guide book. Thanks to Barbara Levy of Barbara Levy Literary Agency and the author Max Arthur, for allowing quotes from the book *Last Post* by Max Arthur, published by Weidenfeld & Nicolson, copyright Max Arthur 2005. Thanks to Sue Robinson for providing quotes from her book *Chitterne – a Wiltshire Village*.

Other sources of reference include The Army Training Estate Public Information Leaflet on ATE Salisbury Plain; *The Story of The Duke of Edinburgh's Royal Regiment (Berkshire and Wiltshire)*, published by The

Regimental Museum, The Wardrobe, Salisbury; *Britain's Aviation Memorials & Mementoes*, by David J. Smith, published by Patrick Stephens Limited; British Military Aviation Timeline pages of the RAF Museum website; *The World War One Source Book*, by Philip J. Haythornthwaite, published by Arms and Armour Press; the website of the Fovant Badges Society; *The Royal Tank Regiment, A Pictorial History 1916-2001*, by George Forty, published by Halsgrove; *A Page at a Time*, by Jack Page, published by Wilton 65; *The history of the Glider Pilot Regiment* by Claude Smith, published by Leo Cooper, Pen & Sword Books; *Politicians at War* by Henry Buckton, published by Leo Cooper, Pen & Sword Books; and *Symbol of Courage, a complete history of the Victoria Cross* by Max Arthur, published by Sidgwick & Jackson.

Many of the photographs in this book have been reproduced from *Hutchinson's Pictorial History of the War*. The drawings, maps and photographs were largely commissioned by Walter Hutchinson, and Virtue and Company Limited, during the progress of the war, and are a unique record of events. They are reproduced with the kind permission of Michael Virtue, Virtue Books Limited, Grindfield Farm, Furners Green, Uckfield, East Sussex.

Thanks also to the following who have helped in some way with information, quotes or photographs: Richard Pittman, Harry Ashford, John Chappell, Evelyn Joan Dancey, Ken Mitchell, Nikki Moore, Ken Don, John Webb, Dave Bamford, Mrs F.M. Tuffs, John Fellows, Mr G. Gilbert, David Piper, John Grosvenor, Geoff Baker, Terry Crawford, Patricia Steel, Barbara Mottershead, Andrew Smith, Phil Williams, Linda Walsh, Brenda Gower, Elsie Lewis, Hazel Green, Robert Brown and Arthur McKechnie.

Thanks also to Geoff White of Geoff White Ltd, and Alix Baker PCAFAS, FRSA, for permission to reproduce their postcards. Thanks to Jim Fuller for permission to reproduce photographs taken and published as postcards by his grandfather Thomas Lionel Fuller of Amesbury, for which he holds the copyright. In connection with the other older postcards reproduced in the book, every effort has been made to trace the copyright holders.

The author is grateful to and acknowledges the following for use of the following images: Geoff White Ltd, 1; Gale & Polden Ltd, 2, 9, 47, 48, 49; A. McKechnie, 3, 7, 8, 12, 23, 24, 25, 27, 28, 29, 30, 33, 40, 44, 58, 87, 88, 97, 98; Kendall of Andover and Tidworth, 4; T.L. Fuller of Amesbury © J.T. Fuller, 6, 22, 32, 36, 67; Kingsway Real Photo Series, 8, 11, 50; The Photochrom Co. Ltd, 10; W.S. Tucker, 12; Mrs F. Tuffs, 13; Terry Crawford, 14, 15, 18, 41, 46; Evelyn Joan Dancey, 17, 38, 68; Bennett of Bulford, 18; Susan Lindsay, Curator, the Museum of Army Flying, 19, 20, 61, 69, 71; GeoGraph/Patricia Steel, 21, 42, 43, 51; Ken Mitchell, 26, 52, 53; Vowles, 27, 33, 44; S.P. Butt, 31; Alix Baker, 35, 37, 70, 74; Australian YMCA series, 39; Q. Erdington & Co. Photographers of Salisbury, 40; Linda Walsh, 45; *Hutchinson's Pictorial History of the War*, with the permission of Michael Virtue, Virtue Books Ltd, frontispiece, 54, 56, 59, 60, 62, 63, 65, 66, 72, 73, 75, 76, 77; Nikki Moore, 57; Sue Robinson/Bert Bailey, 78, 79; Vic Damon/3AD History Foundation, 80, 81; Brenda Gower, 82; Elsie Lewis, 83, 84; Hazel Green, 85, 86; Grant Paton, 89, 90, 91, 99, 100, 101, 102, 105, 106, 107, 110, 111, 118, 119; John Chappell, 92, 93, 94, 95, 96, 103, 104, 109; John Grosvenor, 108; John Webb, 112, 113, 114, 115, 116, 117; Richard Pittman, 120, 121; GeoGraph/Brian Robert Marshall, 122; GeoGraph/Andrew Smith, 123. Illustrations 5, 9, 10, 11, 16, 31, 34, 47, 48, 49, 50, 55, 64 are from the author's collection.

Introduction

Over the years I have often visited Salisbury Plain. At one time, as a member of No. 4626 County of Wiltshire (Aeromedical Evacuation) Squadron of the Royal Auxiliary Air Force, I spent many weekends, and indeed complete weeks, training on the Wiltshire downs. Even though my unit was based at, first, RAF Hullavington and then RAF Lyneham, the downs were an obvious draw when it came to training, and for several years Wiltshire became a second home. Indeed, while training I have probably walked along more of Wiltshire's lanes and chalky tracks than many native-born Wiltshire men. Therefore, when I was asked to write this history of the military involvement with Salisbury Plain, I was not only intrigued to discover more, but felt qualified to do so.

Obviously this is a huge subject, and a book of this size can only scratch the surface. It would be impossible to mention all the units and people who have trained in the area since the land was first purchased towards the end of the 19th century. Instead, we concentrate on the way the area has developed and on the main historical events and changes that have helped shape today's garrison towns. Facts, figures and dates are all very good, but they don't reveal what it was like actually to live and work on the Plain, so where space has permitted I have also included short memories from the soldiers and local residents who have lived through these varied times.

At one point or another in their careers, the vast majority of British service personnel will spend time in the area of Salisbury Plain, as it is now established as the UK's leading military facility. It has seen major changes in the organisation of the Army, witnessed the testing and development of numerous weapon systems, and was the birthplace of Britain's military aviation. Today, its training grounds and military schools prepare our armed forces to serve in an ever-changing and, often, increasingly volatile world. The British armed forces are respected the world over for their training, discipline and professionalism and, for many, Salisbury Plain is where it all begins.

Chapter One

The Early Years

Today, the Army Training Estate (ATE) is controlled from its headquarters at the Land Warfare Centre in Warminster. Around Britain there are 12 regionally based training areas, each of which has its dedicated headquarters and staff. These range from Castlemartin in Pembrokeshire to the Stanford Training Area in Norfolk. But, at more than 94,000 acres, ATE Salisbury Plain is the UK's largest military training area, measuring approximately 25 miles across and ten miles from top to toe. This area includes four towns best described as military garrisons, of which Bulford, Larkhill and Warminster exist alongside civilian settlements sharing the same name, while Tidworth has effectively replaced two rural villages that straddled the Wiltshire/Hampshire border before the arrival of the Army. As well as these towns, the Plain is also surrounded by settlements with strong military connections, including Devizes, Westbury, Amesbury, Andover, Wilton and, of course, the city of Salisbury itself.

Before the Acquisition

With the changing face of warfare at the end of the 19th century, it was critical that both the regular and volunteer forces had access to an area in the south of England in which to practise the military tactics of the age. Large-scale military exercises known as 'manoeuvres' had been held at several locations around the country during the 1870s, and Salisbury Plain seemed ideally suited to join them. It was largely uninhabited and belonged mostly to a small group of rich landowners. It was mainly used for grazing large flocks of sheep, but the shepherds were in the employ of the local manors, so they rarely owned their own properties. Cottages were often tied to the job of their inhabitant. So, in order to acquire large tracts of land on Salisbury Plain, there were relatively few people to coerce into selling up. The situation was conveniently aided by the fact that land prices were depressed at the time. The first land to be purchased was 750 acres at Bulford, bought from a Miss J.M. Seymour of East Knoyle, for the price of £7,500. On this site, acquired on 25 March 1897, Bulford Camp now stands.

1 *Officer of the Royal Wiltshire Yeomanry c.1844, from a watercolour painting by W.R. Younghusband.*

Although Salisbury Plain had been the site of ancient battles, its use as a military training area in the modern sense probably dates from the time of the French Revolution. Fearing internal disquiet and possible invasion, many parts of the country established volunteer cavalry troops known as 'yeomanry'. In the county of Wiltshire, the Regiment of Wiltshire Yeomanry Cavalry was formed on 15 April 1797, and consisted of ten separate troops. One of these troops was raised at Devizes following a meeting at the *Bear Hotel* on 4 June 1794. Another troop was raised at Warminster after a similar meeting in the town on 25 July of the same year, at which a sufficient number of suitable gentlemen volunteered to join. The job of these troops, under the command of Lord Bruce, was to assist local magistrates in keeping order.

In 1798, as the threat of a French invasion looked even more imminent, the yeomanry was joined by companies of volunteers known as 'armed associations'. In a sense these were forerunners of the Home Guard of the Second World War: men willing to protect their local area against foreign aggressors. Again, as the only sizeable town on Salisbury Plain, Warminster was represented by the Warminster Loyal Association. The job of these men would be to protect the town and the countryside within seven miles of it. It's therefore very likely that, by the end of the 18th century, men from both the yeomanry and the armed association would have been observed training and exercising on the western plain. The Loyal Association paraded twice a week for two hours on Monday and Friday evenings, and took part in a monthly field day. It was officered by a captain, lieutenant and ensign, and numbered around eighty, which also included one sergeant major, five sergeants, five corporals, three drummers and three fifers.

Even after the threat of invasion had diminished, the Wiltshire Yeomanry Cavalry continued to exist, and in 1831 King William IV conferred the prefix 'Royal' on them; in April 1863 they were given the additional title of 'Prince of Wales' Own' by Queen Victoria. During the troubled times of the Reform Bill, many rich landowners were given permission to equip their own yeomanry troops in order to quell riots and other disturbances. Thomas Assheton Smith, the owner of the Tedworth Estate at Tidworth, raised a corps of yeomanry cavalry in 1830, which on one occasion was reviewed by the Duke of Wellington at Tedworth Park. Smith, a renowned horseman of the county, took an active role and was granted the rank of captain. His troopers were largely gathered from the tenants of his estate and local farmers, which he mounted and equipped at his own expense.

On 10 December 1859, another public meeting was held at Warminster as once again France threatened to invade. The new force of volunteers raised around the country were the predecessors of the modern Territorial Army. They were known as the Volunteer Rifle Corps, and nine such units were raised in the county of Wiltshire, the 10th Wiltshire Rifle Volunteer Corps being raised at Warminster under the command of the Marquess of Bath, who held the rank of captain. He was supported in this duty by a lieutenant, ensign, assistant surgeon and an honorary chaplain. Later, the nine separate units were amalgamated to form the 1st Wiltshire Volunteer Rifle Corps.

As well as volunteers, many counties had regular regiments affiliated to them, and in 1782 the 62nd (Wiltshire) Regiment was formed. By 1881 the Army had changed again and, under the Cardwell reforms, infantry regiments were amalgamated. The bodies that by then had become the 62nd and the 99th Regiments of Foot became the 1st and 2nd Battalions of The Duke of Edinburgh's

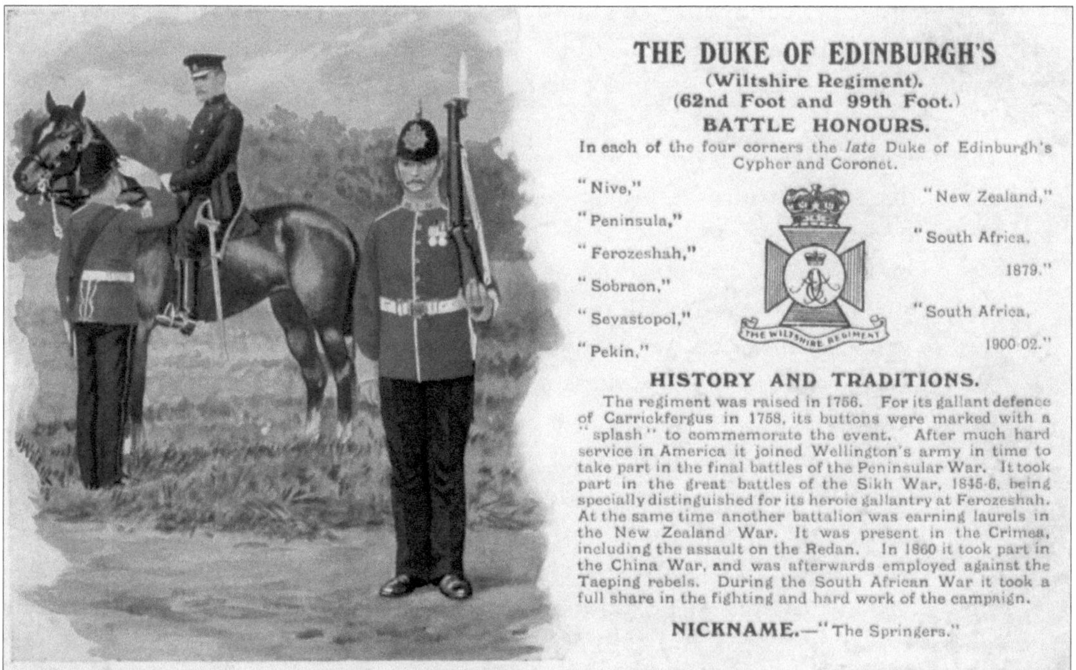

2 *An old postcard showing uniforms of the Duke of Edinburgh's (Wiltshire Regiment) around the turn of the 20th century. The regiment, nicknamed 'The Springers', was formed by an amalgamation of the 62nd and 99th Foot.*

3 *Le Marchant Barracks, Devizes, the regimental headquarters of the Wiltshire Regiment from 1878, when they were built, until 1959. In the centre of this picture, taken around 1905, is the hospital, and to the right some of the married quarters. The hospital was at one time also the sergeants' mess. Today, most of the smaller buildings have gone, while the main barrack block is being converted into flats.*

(Wiltshire) Regiment, with their depot at Le Marchant Barracks in Devizes, close enough to Salisbury Plain to utilise its facilities for training. These barracks, named after Sir John Gaspard Le Marchant, former commander of the 99th Regiment, were completed by 1878, and remained regimental headquarters of the Wiltshire Regiment until 1959. In 1908, after the formation of the territorial force, the 1st Wiltshire Volunteer Rifle Corps became the 4th Battalion of the Wiltshire Regiment.

The first large-scale military manoeuvres on Salisbury Plain were held at Bulford below Beacon Hill in 1872, on land that was still privately owned. These autumn manoeuvres were followed on Thursday 12 September by a royal review, when 30,000 troops either marched or trotted past the Prince of Wales, the Duke of Cambridge and an impressive array of staff officers. Staging these large exercises caused numerous problems which, if they were to be successful, would have to be addressed. Water, accommodation, transport and roads were among the many concerns that caused the government to pass the Military Manoeuvres Act in 1897; the next manoeuvres and review in the area were held at Boscombe Down between 1 and 8 September 1898, and involved more than 50,000 troops. It became increasingly evident that the best way to conduct such manoeuvres in the future was for the War Office actually to own the land, in order that the Army could exercise complete control over it.

Acquisition of the Estate

In 1897 there was a Liberal government in office under the Earl of Rosebery. Lord Lansdowne was the Secretary of State for War, an appointment he had held since 1895. Lansdowne was ultimately responsible for purchasing land for military use, and a month after the acquisition of the acreage at Bulford he set up the War Office Salisbury Plain Committee (WOSP), part of whose brief was to deal with further acquisitions on the eastern side of the Plain and any arising problems. This five-man committee consisted of the Rt Hon. W. St John Brodrick MP, the Parliamentary Under Secretary of State, who acted as president, J. Powell Williams MP, the Finance Secretary, General Sir Redvers H. Buller VC, GCB, GCMG, the Adjutant General, General Sir H. Evelyn Wood VC, GCB, GCMG, the Quartermaster General, and Major General Sir Robert Grant KCB, the Inspector General of Fortifications. By the time this committee was dissolved, on 1 May 1902, it had purchased nearly 42,000 acres of land at a cost of £550,000. Without question the most significant purchase was the 6,618-acre Tidworth Estate from Sir John William Kelk for the cost of £95,000.

Lord Lansdowne wished to disrupt the agricultural population as little as possible. He evidently envisaged a situation in which the Army could use the Plain in harmony with farmers and shepherds. To this end, the landowners were offered the choice of retaining their homes, along with adequate surrounding land, as long as the land in question was utilised according to restrictions placed upon them. Other acquired downland would be subject to restricted grazing rights. As for the shepherds and other farm labourers, the ownership of their cottages had switched from the squires to the War Office, who offered them new rents.

The land affected by these grazing rights was classified as Schedule I, II or III. Under the conditions of Schedule I, full agricultural tenancies were granted on land principally around the training estate's periphery, where livestock could safely graze. The areas were clearly marked, to prevent troops from disturbing them. However, the War Office reserved the right for troops to use these areas whenever failure to do so would create an artificial tactical situation. In this event, compensation would be paid to the farmer for any damage caused. The War Office required that Schedule II land should be kept as permanent grassland, and the Army agreed to provide the tenants with the seed to achieve this goal. Schedule III land covered the remaining estate, where troops could manoeuvre at any time. The farmers were permitted to graze their animals on this land, but the War Office would accept no responsibility for the violent loss of any livestock. However, certainly where Tidworth was concerned, by 1905 there were few remaining local residents. Most farms and cottages at this time were being used to accommodate military or War Office personnel.

4 *This photograph of North Tidworth, taken around 1919, gives us some idea of the village's former rural atmosphere.*

The estate had to be managed like any other vast country estate and, on 19 January 1902, Lansdowne's committee appointed Thomas Stirton as the first of many land agents, with a remuneration of £790 per annum. This sum included his salary, horse allowance, lodging allowance, office expenses and wages for his assistants. He commenced his duties on 11 February, and Ablington House at Figheldean, which has since been demolished, was chosen as his residence and estate office.

Transforming Tidworth

Although the military town of Tidworth stands completely within the county of Wiltshire, it was built on land previously occupied by two small villages, one of which, North Tidworth, was in Wiltshire, while the other, South Tedworth, was in Hampshire. The county boundary actually went through the middle of the military town, a situation unchanged until 1992 when the county boundary was altered. From the outset this established a kind of social snobbery among the soldiers who came to the garrison: South Tedworth was for officers, while North Tidworth was for other ranks. South Tedworth has since been renamed on maps as South Tidworth, although the 'Ted' can still be found in Tedworth House and the Tedworth Hunt.

At the time of the acquisition very little, if anything, remained of South Tidworth, as in the 18th century the owners of Tedworth House had evicted their tenants and taken down their cottages in order to establish their landscaped gardens. Between the two villages, on the Hampshire side of the border, was a hamlet called Hampshire Cross, where it is believed the people of South Tidworth resettled after the evictions. Hampshire Cross stood roughly where Lowa Road can be found in modern-day Tidworth and a number of old thatched buildings can still be seen in this area.

Just to the north-west of Tidworth is Sidbury Hill, which at 735 feet is the highest point on Salisbury Plain. This is an ancient land which in 1897 was largely owned by Sir John William Kelk, whose seat was Tedworth House. His estate included 13 farms, eight farmhouses, 107 cottages and the *Ram Hotel* in North Tidworth. Sir John hadn't been in ownership for long himself, having inherited the estate in 1886 from his father, also called Sir John Kelk. The first Sir John was a self-made civil engineering contractor, who became squire in 1878. Previously Tedworth House had been the home of Thomas Assheton Smith, who died childless in 1858. Smith had been a well-known character in the county, obsessed with hunting and other sporting pursuits. It was he who inaugurated the Tedworth Hunt, and his stables and kennels could reputedly house up to 50 horses and 400 hounds.

Apparently, Kelk had been trying to sell the Tidworth Estate for a number of years. There was little interest, due to the depressed land market, until the Army appeared on the scene in 1897. Although they required the land for training, initially they saw no use for Tedworth House itself, which was one of the grandest buildings in this part of Wiltshire. Having tried unsuccessfully to auction, sell, or let the house, it was decided in 1898 to retain this palatial Gothic mansion as the offices and residence of the garrison's Commanding Royal Engineer, Colonel J.M. Barklie.

So, from the auspicious surroundings of Tedworth House, the building of Tidworth Garrison attached to Southern Military Command was planned and executed. Colonel Barklie moved with his staff into the house, where drawing offices were also furnished in order that the preparation of plans for the new barracks could commence without delay. But, before the Army could move in, a separate horde of navvies was gathered to construct their barracks. These men required accommodation themselves and a shanty town built from corrugated iron sprang up at a location called Brimstone Bottom, which became known as 'Tin Town'. It must have been quite a sizeable concern as in time, as well as huts for labourers, it had its own hospital, mission hall, school, recreation rooms and baths. It was also deemed necessary to build a railway line to service the site. This mini branch line, laid in 1900-1, went from Brimstone Bottom to Ludgershall, where it linked up with the main Swindon to Andover line. Most of the building materials and other supplies that built Tidworth Garrison arrived via the train, and the public was able to use the railway from 1902.

The War Office had purchased in 1897 for £9,350 the necessary land at Ludgershall, which had principally belonged to Manor Farm. This was sometimes referred to as Castle Farm, as it was close to the ruins of an ancient fortification which was known to have been in existence during the reign of King Stephen. Shortly after the purchase Colonel Barklie expressed the opinion that the land could be used to site an Army Ordnance Corps depot and warehouses for the Army Service Corps, as well as railway sidings and lines for distribution. However, besides the subsequent railway, the land was not used for other military purposes until the 1930s.

Before any permanent buildings were constructed at Tidworth, there were already a number of established training camps being used by both regular and volunteer units. By 1903 there were five tented camps around the Tidworth area in regular use during what was regarded as 'the training season'. From north to south, these camps were called Windmill Hill Camp, situated to the west of Ludgershall; Tidworth Pennings Camp, near Sidbury Hill; Perham Down Camp, where permanent barracks would later be built; Tidworth Park Camp, south of Tedworth House; and Park House Camp, south-west of Shipton Bellinger. Interestingly, in the days before Perham Down Camp was upgraded to a permanent camp, a Field Post Office was established there in 1906 to deal with the letters of around 8,000 tented troops.

The first permanent building at Tidworth was an isolation hospital at Brimstone Bottom built in 1900, which was followed in 1902 by Clarendon Terrace, built to house War Office employees who came to work on the site. These were obviously professional men of a status above the labourers at Tin Town, and after the barracks were completed their houses became accommodation for warrant officers.

Between 1904-5, eight barrack blocks were constructed side-by-side below Clarendon Hill. They were each given eastern names, reflecting the Army's heroic association with the British Raj. There were four cavalry barracks, Aliwal, Assaye, Candahar and Mooltan and four infantry barracks, Lucknow, Delhi, Jellalabad and Bhurtpore. The blocks were divided by roads given names like Adampur, Agra, Baroda, Cabul, Dasna, Jamrud and Kirkee. To the north, all of these roads linked up with Bazaar Road which curved around the contour of the hill, or they went south to the Grand Trunk Road. Each barrack block also had a group of associated buildings a short distance away, which included the officers' mess for the battalion accommodated there, and a house for the commanding officer. These houses were named after their respective barracks: Aliwal House, Lucknow House, etc. Most of them were built to the south of the Grand Trunk Road, and there was also a house for the quartermaster of each barrack.

To the north of Bazaar Road, married quarters were built which came to be known as 'Merthyr Tydfils', because they had the appearance of austere Welsh mining houses. In time, over 400 of these would be built below Clarendon Hill. Officers were accommodated to the south of the Grand Trunk Road, on the Hampshire side of the border, establishing the social discrimination between the groups previously mentioned. Each of the officers' messes was also inscribed with eastern names, and in 1912 detached residences were built in Arcot Road for the officers commanding the Cavalry Brigade (Clive House) and the Infantry Brigade (Havelock House).

Troops began to move into the new barracks almost as soon as they were finished, the first contingents arriving as early as 1904. The Cavalry Brigade duly established at Tidworth was the 2nd Cavalry Brigade, commanded by Brigadier General H. de B. De Lisle, CB, DSO, and made up of the 4th Royal Irish Dragoon Guards, the 9th Queen's Royal Lancers, the 18th Royal Hussars and the 2nd Signal Troop, Royal Engineers. The Infantry Brigade was the 7th Infantry Brigade, commanded by Brigadier General F.W.N. McCracken, CB, DSO, which comprised the 3rd Battalion The Worcestershire Regiment, 2nd Battalion The South Lancashire Regiment, 1st Battalion The Wiltshire Regiment and 2nd Battalion The Royal Irish Rifles. In addition to these were No. 20 Company Royal Army Medical Corps, and a section of No. 2 Company Army Ordnance Corps.

5 *Royal Field Artillery Camp at Tidworth Park, around 1903.*

Tedworth House had originally been used as the office of the Commanding Royal Engineer, but in 1905 Lieutenant General Sir Ian Hamilton, KCB, DSO, was appointed General Officer Commanding-in-Chief of Southern Command and Tedworth House became his official residence. From May 1909 it was the residence of his successor, General Sir Charles Douglas, GCB. In 1903 a sewerage works was built at Tidworth Park, and a military cemetery was established on the slope below Sidbury Hill. A veterinary hospital was built in 1904, which was essential at a time when horses were still widely used in the Army. One of the barracks, Delhi, was converted into the garrison hospital in 1907, initially supporting 120 beds, and it remained a military hospital until 1977. Lieutenant Colonel C.E. Faunce was appointed Officer-in-Charge of the Military Hospital, having previously commanded No. 20 Company Royal Army Medical Corps at Bulford from 1904-5. In the area north of the station, towards North Tidworth village, Ordnance Road marks the approach to an extensive area of stores which had developed from around 1903, and in 1910 Ordnance House was built as the official residence of the officer in charge of this depot.

Between Lucknow and Jellalabad was a space that quickly began to fill with social buildings and services. These gave the garrison something of a community feel although, like many military towns, people were constantly coming for short periods of time and going, so it was difficult to create a true sense of community. There was a baker, butcher and small market under a corrugated iron roof, selling everything from sweets, cakes and fish to bicycles, shoes and watches.

In 1907 the first welfare institute for soldiers was built, which in 1908 was followed by a Wesleyan soldiers' home in Bulford Road. This establishment provided the soldiers of the garrison with both social and religious welfare. As well as including a lecture hall and service room, at which the soldiers could

6 *Another view of Tidworth Park Camp during the early part of the 20th century.*

no doubt listen to religious speakers, it contained a concert hall, refreshment room, reading and writing room, and another room set aside for billiards and other games. It was named the Ian Hamilton Wesleyan Soldiers' Home after the Commander-in-Chief of Southern Command. Another welfare institute was built by the Church of England in Kirkee Road and was opened by General Sir Charles Douglas and the Bishop of Salisbury on 27 August 1910.

In 1909 the Army school and garrison theatre went up. The Army school remained separate from the village school until 1980, when the latter was closed down. It had been initially funded by Mrs Assheton Smith, and was built in 1857. By 1880 it had become the North and South Tidworth National School and later the Tidworth Church of England Primary School. When it was closed, in June 1980, all the local children were transferred to the Army school in Bazaar Road, which was still using the same building that had been erected in 1909. The school was then renamed the Tidworth Church of England (Controlled) Primary School.

The Tidworth Garrison Theatre was opened on 4 June 1909 at the junction of Bazaar Road and Lowa Road, with a boxing competition staged by the Royal Fusiliers and a top prize of £10 for the winner. The Pickernell family has maintained close ties with the theatre since it was first opened. Mr Bert Pickernell was the first stage manager, and then manager, having previously worked at the Empire Theatre in Swindon. Bert was succeeded by his son Ken in 1945, who was subsequently succeeded by his own son Tony in 1983, who is still there today (2008). Since opening, the theatre has been used for many different functions, including concerts by orchestras and military bands, regimental dinners, boxing tournaments, variety shows, film shows and lectures. Many famous thespians have trodden its boards, Eric Sykes, Alastair Sim, Will Hay and Norman Wisdom being just a few.

In 1912 the first garrison church was built. Of course, there had always been churches in the Tidworths. North Tidworth, for instance, has a medieval church, Holy Trinity, which was rebuilt in the 15th century, and the rectors retained glebe land on the slope of Clarendon Hill as late as 1846. South Tidworth also had a church, which was demolished along with the rest of the village in the 18th century. The materials from this church were taken to a new site on the slope of Furze Hill, where they were used to put up a mortuary chapel at a new burial ground. However, in 1880 Sir John Kelk built a new church at South Tidworth, St Mary's, which is still there today but rarely visited.

The first garrison church was an iron building opened in January 1909. St Andrew's Presbyterian Church was at the junction with Kirkee Road and St Patrick's Avenue. It closed in 1964. The Church of England garrison church of St Michael was dedicated on 4 May 1912 by the chaplain general in the presence of Lord Haldane, the Secretary of State for War, and the Bishop of Salisbury. Its foundation stone had been laid on 22 May 1911 by Major General Sir Henry

7 *Perham Down Camp, Ludgershall, the only one of the original training camps in the Tidworth area that would be converted from tents to huts.*

Rawlinson, General Officer Commanding-in-Chief of 3rd Division. It was a permanent building of terracotta blocks. Large church parades had been held in the open air at Tidworth Park, but the new church had the capacity to seat 1,200 people. Also in 1912, the Roman Catholic garrison church of St Patrick and St George was built, with a Roman Catholic welfare institute nearby.

In 1913 things really began to improve at Tidworth when a pioneering electricity generating station was built at Candahar Barracks, which was officially opened in 1914. It brought huge advantages to the garrison, one of which was an electric cinema, also built in 1913, which adjoined the Church of England institute in Kirkee Road and was sometimes referred to as the 'Institute Cinema'. But, more importantly, the station supplied power to the ever-increasing lines of military buildings. A row of four houses was built to accommodate the men working on the generating station, and there was a sub-station at Bulford.

After the main construction phase had been completed, the railway station that had been built on the county border to service the site was converted into a proper railway terminal, becoming Tidworth Station. Soon troops were arriving in their hundreds, to march smartly along Station Road to their new homes. The life and appearance of this quiet rural backwater had changed forever.

The Beginnings of Bulford

Although the land at Bulford was purchased before the Tidworth Estate, building work here didn't begin until 1903, and at the beginning it was never on the same scale. There were no brick-built barracks to begin with, just wooden and corrugated iron huts to house the troops. There were some brick buildings, including over 100 married quarters which, following the trend set by their neighbours to the north-east, were christened 'Merthyr Tydfils'. Other permanent brick buildings included riding schools and a gymnasium. From the outset, officers and their wives either found private accommodation in nearby villages, or made do with converted huts. So Bulford must have felt a lot more basic to its early inhabitants than Tidworth.

A total of 1,039 acres was eventually purchased from Miss Seymour, another 1,917 acres from the Bulford Manor Estate of F.C. Ledger Hill, and 2,205 acres at Brigmerston from S.S. Rendell. On the land acquired from Bulford Manor a tented camp was quickly established, and by August 1897 this was already being used by the 4th Cavalry Brigade, which consisted of the Scots Greys, the 1st (King's) Dragoon Guards, the 3rd Dragoon Guards and the 7th Dragoon Guards. By 1900 there was a panoramic spread of white bell-tents, often used by militia, volunteer or yeomanry regiments. An observer noted how most of the soldiers camped here still wore scarlet or blue jackets, although drafts arriving on the Plain to take part in training exercises for the South African War were beginning to be issued with khaki for virtually the first time. So the whole appearance of the British Army was changing to greet a new century, and Salisbury Plain would play a crucial part in these changes.

Early church parades at Bulford Camp would have been held in the open air. All available officers and men would attend and, as well as the sermon preached by the army chaplain, there would be communal hymn singing. These would have been very interesting occasions, as the music would have been provided by the fifes and drums or brass bands of the assembling regiments.

The pre-army village of Bulford was a small settlement on the eastern side of the river Avon, a watercourse that followed a roughly straight line across the Plain from north to south. In the 1901 census, the inhabitants of the community numbered 343. The new Army camp was built to the east of this village, while another settlement called Durrington sat on the western side of the river. There was plentiful evidence on the downs of Neolithic and Bronze-Age man, particularly the barrows and field systems that can be discerned in the gently undulating landscape.

In similar manner to Tidworth, a Tin Town sprang up to accommodate the navvies working on the construction of the camp. As the building work at

8 *The garrison theatre at Tidworth, which was opened on 4 June 1909 with a boxing competition staged by the Royal Fusiliers, with a top prize of £10 for the winner.*

9 The remains of a postcard showing Water Street, Bulford. This photograph, probably taken before the First World War, shows soldiers and local people mixing together.

Bulford was not as ambitious, the shanty town would have been a lot smaller, and in the same census only 45 construction workers are listed. It's interesting to note that 608 soldiers are also included, the vast majority of which would still have been living in tented accommodation. Shopkeepers and traders set up their businesses initially at Tin Town, with access off the suitably named Bond Street. The shrewd landlord of the *Rose and Crown* in the High Street had rebuilt the pub in 1896, just in time for this influx of thirsty soldiers and construction workers to appear. Again like Tidworth, Bulford's Tin Town included a large open-fronted shed made of corrugated iron. This contained the garrison market, which was eventually pulled down in 1969 and turned into a car park. There was also a fire station, again manufactured from corrugated iron, the exact date of which is unknown, but its design would suggest it certainly existed before the First World War.

By 1903 two separate camps had been established to the east of Bulford village. Bulford Camp, the farthest away, was still largely tented while, as its name suggests, Bulford Hut Barracks, was a more permanent establishment. The barracks, divided into eight lines, were separated by three roads, which corresponded roughly with Haig Road, Wing Road and Rawlinson Road in modern-day Bulford. Seven of the eight lines were identified by the letters A to G, while the eighth, occupied by the Army Service Corps, was simply known as the ASC Lines. Various brigades of the Royal Field Artillery occupied A, B, C and D Lines, while the Royal Engineers could be found at E Lines. It's difficult to find out exactly who was in occupation of F and G Lines, so they were probably being used by cadres of other supporting service arms, or anyone else who helped the garrison to function. For instance, it is known that the infirmary stables could be found in F Lines. And, although there was no Church of England garrison church until after the First World War, there was a Roman Catholic church situated in G Lines from October 1910 until April 1912. In June of the same year, another Roman Catholic church opened in F Lines, which subsequently closed in March 1915.

Each of these separate lines was a complete unit, with barracks, guardroom, regimental offices, messes for both officers and sergeants, and married quarters for officers and all other ranks. There were laundries and two schools, one in A Lines and the other in B Lines. The barracks, constructed between 1900 and 1901, had wooden frames which were then covered with felt and corrugated iron. Each barrack could house between ten and twelve men, and was lit by paraffin lamps and heated by a single stove. For the warrant officers several two-storeyed brick houses were built, with slated roofs. A supply of electricity would eventually arrive at Bulford in August 1914, after the building of the generating station at Tidworth. All of this must have seemed like extreme luxury to the soldiers at Bulford Camp, still living under canvas, which had to be broken up in November 1901 after several men died of pneumonia.

By 1910, F Lines had been re-appropriated as married quarters. Also around this time, the Bulford 'Merthyr Tydfils' were built in Vimy Crescent, adjoining the south-west boundary of the barrack area. There were five blocks in all, with 12 terraced houses in each, and all built to a very similar design to those at Tidworth. And in what would later become Mons Avenue, two further rows of houses were built, one of which consisted of five dwellings and the other twenty.

In order to move building materials, supplies and troops, between 1903-6 a railway network was built from Amesbury to Bulford Camp (the Amesbury and Military Camp Light Railway). There was a terminus at the camp supply depot

10 *The sprawling tents of Bulford Camp pictured around 1914.*

11 *B Lines Bulford Camp.*

and two stations for military personnel, one at the bottom of Horne Road, and another at Sling. The former has now been demolished but the Sling platform can still be seen to the south of the old Sling pumping station. The villagers benefited, with a public station to the south of the village, and the southern end of the High Street was renamed Station Road. Another spur would eventually run off the Amesbury-Bulford line to serve Larkhill. This in turn threw off a series of further spurs and branch lines, serving various other camps, and in 1910 the Army built its own linking road between Bulford and Tidworth.

As the new barracks at Bulford became occupied, there was a need to provide recreational and welfare services for the off-duty soldiers. A Wesleyan soldiers' home was opened in 1903 on a site next to land on which St George's Church would be built after the First World War. Also in 1903, a Church of England soldiers' institute was officially opened by Lieutenant Colonel Stirling, supported by Lady Wimbourne, who had laid the foundation stone in October 1902. At a cost of £7,000, this institute was built to the east of St George's Church, adjoining the garrison theatre, which was itself the first place of entertainment to be provided at the garrison. It is not clear when this was built, or how long it functioned as a theatre, but by 1939 it was used almost exclusively for screening films.

A third welfare home was established by the Perks sisters, two ladies who began their welfare work in Winchester around 1875, firstly providing facilities to the militia during their annual camp, and then to regulars. They founded both a soldiers' home and mission halls around Winchester, and since at least 1896 there had been tents from their Winchester Soldiers' Home on Salisbury Plain to provide help for the troops training there. The *Hampshire Observer* mentioned the laying on 11 October 1902 of the memorial stone of the new home, which fully opened on 4 June 1903. Its architect was Thomas Stopher of Winchester and it was built by Messrs Wort and Way of Salisbury. It stood where the Roman Catholic garrison church now stands, and the avenue of trees that leads to the church formerly led soldiers to their home. The site was apparently obtained through the good offices of Lord Roberts.

Although Bulford Hut Barracks were largely inhabited by gunners, Bulford Camp was mainly used by cavalry and yeomanry regiments at its inauguration; the men lived in tents but several brick riding schools were built. Remnants of at least three of these were later employed for other purposes. In 1900 Bulford Camp was commanded by Brigadier General V. Hutton CB, and the example of the Boer guerrillas during the South African War and their mobility in the field led to the recognition that mounted infantry could be used with great effect. To this end, three schools of instruction for mounted infantry were established around the country, one of which was at Bulford. In May 1903 a mounted infantry battalion was formed at Bulford Camp, but the school itself only lasted until July 1906, after which it was temporarily used by the Cavalry School from Netheravon. The Cavalry School at Netheravon was established in the summer of 1904, at Netheravon House, but after its temporary move to Bulford, in January 1906, it returned to its former home the following summer.

Two of the remaining riding schools at Bulford can be found at the junction of Mons Avenue and Haig Road, one of which was later occupied by the Bulford Study Centre while the other was employed as a NAAFI messing store. The third riding school, at the corner of Rawlinson Road and Provost Road, has enjoyed varied employment. It has been a study centre and the gymnasium of the United

12 *A Wiltshire Regiment Church Parade on Salisbury Plain in July 1913. Before garrison churches were built, church parades were held in the open air. All available officers and men would attend, and the music would be provided by the band of the regiments gathered. Here the band of the Wiltshire's is clearly seen.*

13 *This photograph was taken somewhere on Salisbury Plain in 1905, and is titled 'The Mixed Squad'. It is obviously taken at one of the earliest brick-built rifle ranges, probably at Bulford, and shows soldiers of the United States Army posing with British troops represented by several regiments including the Royal Field Artillery, the Hampshire Regiment, the South Lancashire Regiment and the Wiltshire Regiment.*

Kingdom Land Forces School of Physical Training. It was also used as a concert hall by the Entertainment's National Services Association (ENSA) during the Second World War.

In 1907 the 3rd Division moved into Bulford Camp under the command of Major General W.E. Franklyn CB, and remained there until the start of the war. By this time, of course, Bulford Hut Barracks had become established as the Royal Artillery's principal base. Field guns were still drawn by teams of horses so stabling was also of paramount importance, and around 1910 permanent red brick barracks were built either side of the Marlborough Road, which were quickly followed by permanent shops and banks. Bulford was starting to have a more permanent feel about it. A 200-bed hospital was established in 1904, which preceded the garrison hospital at Tidworth. It was under the charge of Lieutenant Colonel Faunce who, as previously mentioned, commanded No. 20 Company Royal Army Medical Corps before moving to Tidworth Military Hospital.

Since the days of Thomas Assheton Smith, Salisbury Plain had been regarded as fine foxhunting country, and indeed, even after the acquisition of the Tidworth Estate, the Tedworth Hunt continued to exist. In 1907 the Master of the Hunt was Captain A.E. Hussey, who was approached in that year by a group of officers from Bulford who had the idea of forming their own hunt to course the many hares that abounded on the Plain, thus avoiding a conflict over quarry. Hussey agreed, and the Royal Artillery (Bulford) Harriers were formed. At the start their kennels were situated to the south of Bulford Hut Barracks, where Bulford Road met Milne Road, and were a great success. However, because the artillery brigades were away on active service during the First World War, the painful decision was taken in 1917 to put down the pack of harriers.

Land for Ranges

It would of course be wrong to think that the War Office purchased the whole of Salisbury Plain. Indeed, driving along the A303 today you see little evidence of the military presence as your route is lined by productive farmland. In the early years, when the plan was to buy up land in the eastern Plain, the one major obstacle was the Coombe Estate, which extended from Coombe towards Everleigh. As training commenced in earnest the Coombe Estate often presented an unwelcome obstacle hampering north-south movement on tactical exercises.

Sources agree that when the War Office first decided to purchase land on Salisbury Plain it was to accommodate exercises for both regular and volunteer troops. There was no intention to build permanent camps, and troops would be accommodated in tents during the period of their manoeuvres. So by 1900, and before any permanent building programme was under way, troops were already manoeuvring on the Plain from these tented camps. By 1903 temporary summer camps were taking place at Windmill Hill, Tidworth Pennings, Perham Down, Tidworth Park and Park House.

As well as being tasked with purchasing land, the War Office Salisbury Plain Committee was concerned with its subsequent use. One of the initial problems of establishing permanent camps was the lack of suitable drinking water. The water from the rivers Avon and Bourne and other sources on Salisbury Plain was considered suitable for horses to drink, but not for the men, so, even while on exercise, provision would have to be made for bringing an adequate supply on to the downs. The Committee members discussed this at length, and the Inspector General of Fortifications made the point that men on manoeuvres would almost certainly drink river water, as they would on active service. Thirsty soldiers in the field would probably drink from any natural fresh water source available to them, so this shouldn't cause a hindrance.

Although the principal purpose for purchasing the land was in order for troops to exercise, Lord Lansdowne was keen on building ranges at which the men could shoot their rifles. The Adjutant General was opposed to the idea, wanting the training area to be used for exercising the men in battle tactics, under which circumstances there was no need even to fire weapons. In fact, while a regiment was using the facility he considered it a waste of their time if they were distracted by shooting practice. He also suggested that if, they had to shoot, they could do it without ranges on Salisbury Plain, following the practice observed in India. However, around the turn of the century, several rifle developments had produced the Short Magazine Lee-Enfield, the range of which was far greater than weapons of preceding decades. Concerns were raised about the safety of existing ranges and, sympathetic to these concerns, Lord Lansdowne envisaged Salisbury Plain as the perfect place to practise with new calibre, longer range weapons.

The building of both the first rifle range and the first artillery range began in 1898, on land to the west of the river Avon. The first temporary rifle range was built on the downs to the west of the village of Netheravon, near Newfoundland Farm, whose inhabitants were concerned by the fact that the range was directed straight into the farm buildings. Unfortunately they had little say in the matter as the farm, which no longer exists but was situated near Old Farm Clump, now belonged to the military. By way of compensation the farm rent was reduced to ten shillings a week, but when the range was required for practice the inhabitants were obliged to vacate their home. The present ranges west of Netheravon date from the First World War, when the Machine Gun School moved into buildings vacated by the old Cavalry School.

The Netheravon Estate had been purchased from Sir Michael Hicks Beach in 1898. The War Office acquired his 7,813 acres for £93,411. Considering the amounts paid to other local landowners, this was a very good price, which caused public disquiet at the time because Sir Michael, who had inherited the land through his wife Henrietta Beach, was Chancellor of the Exchequer. In 1904 the Cavalry School was established at Netheravon House, with its indoor riding school, barracks, houses, villas for officers, and larger villa for the commanding officer all built in the grounds of the estate.

The first permanent rifle ranges were built along the Bulford to Tidworth road shortly after the construction of the camps began. The old Marlborough road, which ran northwards from Sling to Everleigh, passed through the danger area as it by-passed Tidworth and so had to be closed to public transport, and this was one of the considerations when the Army built the military road connecting Bulford to Tidworth. By 1900, 15 permanent rifle ranges had been built in four localities. There were five ranges to the east of Bulford Camp, named 'A Range' to 'E Range', 'A' being closest to the camp. The other ranges were built at Larkhill, Tidworth and Perham Down.

Developments in artillery also required large tracts of land. Guns were becoming more mobile and versatile, and able to fire over longer distances. A new form of artillery warfare was termed 'fire and movement', and in order to become proficient in such techniques the various artillery brigades needed room to manoeuvre. Suitable locations were tested on Dartmoor and at Aldershot, Hay and Romney, before an area on Salisbury Plain to the west of the Avon was used in August 1899. Bulford Camp may well have been the Royal Artillery's principal base, but Larkhill was soon to become established as the artillery training area. Here, just to the north of Stonehenge, were vast expanses of open downland, suitable for practising with the long-range guns of the day.

The artillery ranges, in part of what is now the Larkhill Impact Area, came into use for the first time in 1899, when several batteries came down for practice camp. Their stay was short, from 10 to 24 August only, and firing was restricted to the interval between nine and two o'clock to interfere as little as possible with agricultural activities. Two years later the range was extended

14 *The Honourable Artillery Company Infantry parade at Bulford in 1910 wearing bearskins. The sender of the postcard says, 'We have to keep on practising for Friday. I do not know yet where we are going to stand.' This refers to a detachment of HAC Infantry attending the funeral of Edward VII on 20 May. It is believed that on the day they stood at Whitehall with the Brigade of Guards.*

to the old Salisbury-Devizes road and in 1911 a new range was acquired at West Down.

Each year batteries visited the Plain for short summer stays to fire their practices. There were no permanent buildings at Larkhill until 1914, so the troops had to live in official tented camps during their stay. And it was a short stay at that, during which they merely fired their guns and departed, taking no further advantage of the training area. It was not until 1914 that the tents were eventually replaced by hutted camps, or military activities took place all year round. The two ranges being used at the time for these practices were called 'Larkhill Artillery Range' and 'Rollestone Artillery Range', which was later renamed 'West Down Range'.

Initially there were three official camp sites accommodating visiting gunners: North Camp, South Camp and the Royal Field Artillery Camp. North Camp was near Tilshead, to the north-east of the present-day West Down Camp. South Camp was a short distance to the east. Both of these were also used by cavalry regiments, from time to time, who had been equipped with 'pom poms', or 3.7mm Maxim heavy machine guns. The Royal Field Artillery Camp was the first to be situated on land that would later become Larkhill Garrison, and it was roughly situated near the garrison church.

15 *In the Baths at Larkhill, 1908. Many (if not all) of the pre-First World War camping sites had open-air baths such as these. The water wasn't changed that often, and so the use of soap was forbidden because it would cause scum to form.*

Further camps were soon to follow, such as Pond Farm Camp on West Down and Lark Hill Camp, which straddled the Packway to the west of the Royal Field Artillery Camp, later renamed Durrington Camp. The Packway was once a dusty downland track that now forms the main road through the heart of Larkhill. Further west still was Fargo Down Camp, then Rollestone Camp, Bustard Camp and Hamilton Camp, the latter named after Sir Ian Hamilton. At the western end of Hamilton Camp, an ammunition depot was later built, subsequently replaced by a larger depot to the south. If these tented camps felt inhospitable to cold, wet, and miserable soldiers, *The Bustard Inn*, situated at the remote centre of the western Plain, must surely have felt like an oasis.

Around 1910, some brigades of Royal Field Artillery based at Bulford Hut Barracks would no doubt have trained on Salisbury Plain but, strangely, most gunners from Bulford rarely practised at Larkhill, instead going to Okehampton in Devon or Trawsfynydd in North Wales. The ranges at Larkhill were predominantly used by gunner brigades from Aldershot or by volunteer

16 *The 6th Hampshire Regiment arrive on Salisbury Plain in 1909 at Ludgershall Station. This postcard was postmarked from Ludgershall and the sender writes, 'Can you find me in this picture? Having some rotten weather. Been raining all the morning. Have been having a lecture in our tents, awaiting for dinner now.'*

regiments. The training season lasted from May to the middle of September each year, after which the downs fell eerily silent.

Since April 1859 the School of Gunnery had been situated at Shoeburyness, but the advantage of moving it to Larkhill was becoming increasingly apparent. During the summer of 1912 it was attached to Larkhill for the season, proving so successful that in 1914 the decision was taken to move it permanently to Wiltshire. Of course, in 1912 tents still lined the downs, but in 1914 Larkhill was about to undergo a massive change.

So Larkhill slowly emerged as the third military town on Salisbury Plain, with traders flocking to an area below the garrison church. Clustered here in wooden and tin huts, shops and businesses associated with an emerging garrison town must have given the place the feel and appearance of a Wild West frontier settlement in the heart of the English countryside.

17 *C Squadron, Royal Wiltshire Yeomanry, camped at Tisbury in May 1914. This photograph, provided by Evelyn Dancey, shows her father Hubert James Cove in the middle.*

18 *Gamma Army airship at Larkhill, c.1911.*

Birth of the Royal Flying Corps

Salisbury Plain has an important place in the history of British military aviation. In its infancy, long before the Royal Air Force had been envisaged, aviation came under the control of the Army. It was at Larkhill that the very first military aerodrome in Britain was founded. In 1909, a gentleman called Horatio Barber, an early aviation enthusiast, rented some land at Larkhill and built a shed in which to keep his brand new aeroplane.

At nearby Bulford, a certain Captain J.D.B. Fulton of the 8th (Howitzer) Brigade, Royal Field Artillery, was also experimenting with powered flight, having purchased a Bleriot monoplane. Fulton and another enthusiast, Mr G.B. Cockburn, stabled their own aircraft at Larkhill and built sheds next to Barber's in which to keep them. In May 1909 the War Office ordered the construction of two further buildings on the site to be used by both the Army and the Navy for aviation training purposes. When Barber left, in October 1910, the War Office took control of his shed, which was thereafter used by their flying school. Here, aspiring aviators qualified for the Royal Aero Club licence, similar to those awarded at Brooklands and Hendon. G.B. Cockburn, although a civilian, is reputed to have taught the first four Royal Navy pilots to fly.

In June 1910, the British and Colonial Aeroplane Company, who would later become the Bristol Aeroplane Company, asked permission to establish their flying school at Larkhill. Permission was granted, and they proceeded initially to build a three-bay hangar. Some of their structures, still visible in Wood Road, are quite possibly the oldest buildings in Britain connected with aviation. By the end of 1910, Larkhill had already become a busy centre for Britain's fledgling interest in military aviation.

Balloons had been used by the Royal Engineers during military operations towards the end of the 19th century, in the Sudan, Bechuanaland and South Africa. Some people envisaged aircraft being used for similar purposes, notably as an aid to reconnaissance, for observing troop movements or, perhaps most importantly, as a way of directing gunfire onto enemy positions. During the late

1890s, Salisbury Plain had been the venue for several courses covering the use of balloons. These were attended by officers of all arms, but were particularly aimed at gunners. In May 1909, further experiments were held when the 12th (Howitzer) Brigade, Royal Field Artillery, exercised with No. 2 Balloon Company of the Royal Engineers. For the duration of these trials, the Balloon Company camped at Durrington and the gunners at Larkhill.

In 1911 the Royal Engineers formed an air battalion with its headquarters at Farnborough. The battalion was split into two companies: No. 1 Company, also based at Farnborough, was equipped with airships, kites and balloons, while No. 2 (Aeroplane) Company at Larkhill, under the command of Captain Fulton following his transfer into the Royal Engineers, was concerned with the military use of powered aircraft.

The importance of aircraft in warfare soon became evident, and on 13 April 1912 the Royal Flying Corps was formed, with both a military and naval wing. However, in 1913 the naval side branched off to form the Royal Naval Air Service. No. 2 (Aeroplane) Company of the Royal Engineers Air Battalion subsequently became No. 3 Squadron, Royal Flying Corps, under the command of Major Brooke-Popham. Larkhill's No. 3 Squadron was, therefore, the very first unit in British military circles to be equipped with aircraft. The motto of their descendants, No. 3 (Fighter) Squadron of the Royal Air Force, is, quite appropriately, 'The Third shall be First', or 'Tertius Primus Erit'.

A Central Flying School was opened at Upavon on 19 June 1912, just to the north of Larkhill, under the command of Captain Godfrey Paine, MVO, RN (later Rear-Admiral), while an aircraft factory was established at Farnborough.

19 *Army and naval personnel in front of aircraft and hangars at the Central Flying School, Upavon, in 1912.*

Further members of staff and flying instructors were soon appointed to the flying school, among them Captain Fulton, who would later rise to the position of Chief Inspector of the Aeronautical Inspection Department, but died tragically young in November 1915.

Salisbury Plain is, of course, strewn with ancient earthworks, and when the airfield was created on Upavon Down archaeologists were invited to watch the levelling of a Bronze-Age barrow which was obviously regarded as a hazard to aviators landing and taking off. The first 800 acres at Upavon, to the south of the Iron-Age settlement known as Casterley Camp, were purchased in 1898 for use as an Army firing range. This was followed a few years later by the purchase of 425 acres on Upavon Down to be used by the Central Flying School. The decision to locate an airfield at Upavon wasn't without its critics, as the site was very exposed and sloped in a south-westerly direction from 580ft above sea level at one end to 480ft at the other. But it wasn't long before several hangars and associated buildings had been constructed about a mile to the east of Upavon village, and the airfield duly established.

In August 1912 the first British military aeroplane trials were conducted at Larkhill. Twenty-four aircraft were entered and 16 new hangars were built to accommodate them, although after the completion of the trials they were taken down and removed to Farnborough. The American aviation pioneer Samuel Franklin Cody won the speed competition in his Cody Cathedral aeroplane but he was killed a year later while landing on Laffan's Plain, Farnborough.

On 13 August 1912, Major Hugh Trenchard, who became known as the father of the Royal Air Force, was posted to the Central Flying School as an instructor after winning his aviation certificate at Larkhill. He was Assistant Commandant from 1 October 1912 until 7 August 1914. Today, the former station headquarters at Upavon, built in 1914, has been turned into a museum dedicated to Lord Trenchard, who apparently watched his pilots training from its verandah. Another early instructor at the flying school was Captain John Salmond who, like Sir Hugh Trenchard, went on to become Marshal of the Royal Air Force.

Salisbury Plain remained at the forefront of aviation experimentation for the next few years. In June 1912 a British altitude record was set by a Royal Aircraft Factory BE2 observation aircraft when it flew over Salisbury Plain at 10,561 feet. Between 16 and 17 April 1913, Lieutenant R. Cholmondeley of No. 3 Squadron, Royal Flying Corps, flew a Maurice Farman biplane from Larkhill to Upavon and back on what was the first ever night flight made by an aeroplane.

With all this aerial activity going on, Salisbury Plain was the location for a number of early aviation accidents and tragedies. At a crossroads two miles east of Shrewton, known as Airman's Corner, a stone cross stands to the memory of Captain E.B. Lorraine and Staff Sergeant R.H.V. Wilson, both of the Royal Engineers Air Battalion, who met with an accident near the spot on 5 July 1912.

20 *Aeroplanes on parade for an inspection by Lord Roberts at Netheravon on 29 June 1914.*

They were killed when the Nieuport monoplane in which they were flying crashed a mile and a half west of Stonehenge in what is the first recorded fatal flying accident on Salisbury Plain.

Also near Stonehenge, in the south-east corner of Fargo Plantation, is a cross commemorating Major Alexander William Hewetson, an officer of the 66th Battery, Royal Field Artillery, and one of the earliest members of the Royal Aero Club. Hewetson was killed on 17 July 1913 flying a Bristol Coanda two-seater monoplane from Larkhill while taking the test for his aviator's certificate. The crash site on the former aerodrome is marked by a fenced-in stone slab resembling a grave. And on 3 October 1913, Major G.C. Merrick became the first pilot to be killed while training at the Central Flying School when the Short biplane in which he was flying crashed. Many more tragic deaths would follow.

The Royal Flying Corps took part in their very first review at Perham Down on 22 May 1913, when General Sir Horace Smith-Dorrien, GCB, GCMG, DSO, reviewed some 8,000 troops. The appearance of aircraft in the sky overhead would undoubtedly have been the highlight of the occasion.

The War Office also decided to build an airfield at Netheravon which was split over two locations known as 'Upper' and 'Lower', 'Upper' being the smaller of the two. Hangars were built between 1912 and 1913 and No. 3 Squadron moved there from Larkhill on 16 June 1913, and was joined by No. 4 Squadron from Farnborough a few days later. Before suitable accommodation had been provided for the men of No. 3 Squadron, the officers had been quartered at *The Bustard Inn*, while other ranks lived under canvas at Bulford Camp. For the final period before their new home was made available, they were accommodated in the cavalry barracks at Netheravon.

In June 1914 the Bristol Flying School left Larkhill and moved to Brooklands, leaving the aerodrome completely at War Office disposal. This proved fortuitous as, also in 1914, the military wing of the Royal Flying Corps gathered in Netheravon for what became known as the 'Concentration Camp'. Five squadrons, Nos 2 to 6, all gathered together, with visiting personnel living under canvas. During the course of the concentration camp the pilots took part in discussions and lectures on the principles and operational techniques of the world's first effective air force. They also took part in exercises on methods of observation and reconnaissance, experimented with photography and wireless telegraphy, and flew simulated offensive actions, which included directing artillery fire onto enemy positions.

Shortly after the concentration camp had finished, war was declared and Netheravon, and the Central Flying School at Upavon, were poised to become both the main training bases for new pilots, as well as the forming-up locations for squadrons preparing to depart for the Western Front.

21 The Bustard Inn *is one of a number of public houses that have been familiar to service personnel ever since the War Office first purchased land on Salisbury Plain. In 1913, the officers of No. 3 Squadron Royal Flying Corps were billeted here before moving into new accommodation at Netheravon.*

Chapter Two

The First World War

By the time war was declared on 4 August 1914, Salisbury Plain was ready to play an important role. By now its permanent barracks and camps were already inhabited by a diverse group of soldiers: the 3rd Division were at Bulford Camp, with the Royal Field Artillery and Royal Engineers providing the major components at Bulford Hut Barracks; at Tidworth the 2nd Cavalry Brigade and 7th Infantry Brigade had become well established and settled into routine. The Royal Flying Corps continued to grow from several locations, adding a new dimension to modern warfare. On 4 August itself, while the 4th Royal Irish Dragoon Guards and the 9th Queen's Royal Lancers were about to mount a tent-pegging display at the Andover Flower Show, news reached them of the outbreak of war and they galloped back to their barracks with all possible haste.

Flower shows and peacetime soldiering were now things of the past. Everything was about to change as the vastly increased wartime army had a suitable area in the south of England in which to prepare volunteer and conscripted soldiers for deployment to wherever British troops were engaged around the world. Over the next few years serried ranks of white bell-tents appeared all over the downs, the permanent barracks built at Tidworth and Bulford being unable to cope with the influx of troops. So a massive building programme was soon under way in order to provide accommodation, not merely for every branch of the British Army, but for men from the Empire and Dominions, such as Canadians, Australians and New Zealanders.

Housing a Wartime Army

The most significant changes took place at Larkhill where, before the war, there were no permanent barracks at all. A Royal Engineer officer, Major B.H.O. Armstrong, his chief assistant Mr J.D. Michel, and his staff of draughtsmen swiftly designed a sectional hut which could be erected very quickly using local labour, which became known as the 'Armstrong Hut'. Entire camps of these huts were built to accommodate wartime soldiers at both Tidworth and Bulford,

22 *This postcard, sent from Bulford Camp in August 1914, shows various military scenes on Salisbury Plain.*

while at Larkhill by December 1914 some 34 of these battalion-sized huts had been built, soon to be occupied by units of practically every arm of the British Army. This expansion led to the disappearance of the pre-war aerodrome, the sheds of which are still in evidence in Wood Road and were subsequently used for other purposes. Most of these camps were only temporary and have since been demolished, although married quarters were later built on many of the sites using the facilities that had been put in place during the war, including drainage, electricity and water supplies.

Although Major Armstrong had designed the new huts, the contract for building the camps was awarded to one of the biggest engineering contractors of the age, Sir John Jackson Ltd. Jackson's company gained notoriety as the company responsible for the devastation of the little fishing village of Hallsands on the south Devon coast after the Admiralty employed it to enlarge the docks at Devonport. Jackson did this by laying down huge amounts of concrete, for which sand and shingle were required, an abundance sitting conveniently off the coast of south Devon. For centuries Hallsands had been protected by a shingle ridge but, after the dredging and its reduction in size, the village became vulnerable. When the first major storm occurred in 1917 the community was quite literally swept into the sea.

Jackson was also given the contract to lay the railway system that would initially benefit the building of Larkhill and, later, the soldiers encamped there. The Larkhill Military Railway, built between 1914 and 1915, went from a place called Ratfyn, about 700 yards north of Amesbury station, to Larkhill, with branches to Rollestone Camp and the airfields at Stonehenge and Lake Down, both of which were opened in 1917.

Bulford Camp, which was still under canvas in 1914, would also increase in size with the building of new hutted camps. The first of these was Sling Camp, which was built on a site between Sling Plantation and Gaza Road. Sling was intended to be used by the New Zealand Expeditionary Force once it arrived in Britain from the southern hemisphere. In preparation for the arrival of their comrades, New Zealand soldiers already stationed in Britain set about helping the contractors to build the camp. The general area today is easily identified by road names such as Auckland Road, Wellington Road and Kiwi Road. However, the New Zealand Army didn't come directly to Britain as expected, landing instead at Alexandria in Egypt, ready for deployment against the Turks in the Dardanelles. Because of this, the troops helping to build Sling were ordered to join them, and the camp was finished with the aid of men from the Canadian Expeditionary Force. Soldiers from New Zealand did eventually arrive at Sling, which afterwards became known as Anzac Camp and was mainly divided into four areas: Auckland Lines, Wellington Lines, Otago Lines and Canterbury Lines.

Of the early tented campsites around Tidworth, only Perham Down was chosen to be converted to huts. The hutted camp here was built in 1915, and in 1916 it became No. 1 Training Group for the 3rd Australian Division. In time, four Australian command depots were established around the country, where soldiers released from military hospitals could prepare for their return to front-line duties, Perham Down being chosen as the location for No. 1 Command Depot. King George V visited Perham Down on 8 November 1917, arriving at Ludgershall on the royal train and then riding to the camp on his horse, where he carried out an inspection of troops.

The surrounding countryside was also strewn with wartime camps, close enough to enjoy the training facilities that the Plain offered. There are far too many to list, but a few would include Tisbury, Sutton Mandeville, Compton Chamberlayne and Fovant on Cranborne Chase, where there were transit and training camps. Fovant is particularly worthy of mention because of the series of badges carved out of the chalk on the hills above and still much in evidence today. The villages and fields at Fovant, Sutton Mandeville and Compton Chamberlayne were swamped by a mass of military buildings, complete with barracks, hospital, parade areas, rifle ranges, camp cinema and YMCA huts. A military railway laid to serve the camps branched off the main London line from Salisbury. Thousands of men from all parts of Britain and the Empire stayed at these camps, either before their departure for the Western Front, or on their return from it. At Fovant and other local churchyards there are rows of war graves belonging to soldiers who died whilst here. Many regiments began to carve badges on the hillsides, possibly in remembrance of fallen comrades, and witness-accounts prove that

23 *Sandhill Camp, Longbridge Deverill.*

some of these were in existence as early as 1916. By the end of the war there were at least 20 discernible examples, but many no longer survive.

The main badges on view at Fovant today are as follows: the Royal Wiltshire Yeomanry, the 6th Battalion The City of London Regiment, Australian Imperial Force Badge, Royal Corps of Signals, the Wiltshire Regiment, the London Rifle Brigade, the Post Office Rifles, and the Devonshire Regiment. Above Sutton Mandeville there were further badges, including the 7th Battalion The City of London Regiment and the Royal Warwickshire Regiment.

The area around Warminster was particularly blessed with wartime camps. The villages of Codford St Mary, Codford St Peter, Sutton Veny and Heytesbury were among the rural outposts that became temporary homes to thousands of men from around the world. When war was declared, yeomanry troops from Warminster, Salisbury and Rood Ashton were billeted around Warminster, the allocation of horses taking place in the Portway. In time, the 26th Division concentrated around the town, some billeted in local homes, with others no doubt under canvas on the downs and many camped at the Codfords. In Station Road, the Army Service Corps established a depot and while there, in August 1915, the 26th Divisional Train, Army Service Corps, was known to have held regimental sports on ground to the north of Cop Heath Lane.

For a small rural village, about two miles to the south-east of Warminster, Sutton Veny was to experience a major facelift during the course of the war. In fact its location made it a suitable place at which to station troops. By November 1914, a major programme was under way to erect several hutted camps over a large area of Sutton Common. There were at least 15 temporary camps, which could be found at several locations including Cooper's Bottom, while others were located in the Sandhill area between Southleigh Wood and Longbridge Deverill. The materials to build these huts were ferried to the sites by the Sutton Veny Camp Railway, which

24 *Temporary camp at Corton.*

went from Heytesbury station to Sandhill. The massive building programme continued until the spring of 1915, with the first troops, a contingent of the 26th Division, moving in on 28 April 1915, although many of the camps still remained unfinished. Before the 26th Division moved in, one of the camps at Cooper's Bottom had been used to house internees, German or other enemy nationals living in the country at the start of the war who were considered a threat to national security. Prisoners of war arriving from the continent around 1916 were also interned in the camp. After the war the huts were removed, the camp railway was lifted, and Sutton Veny returned to peacetime conditions.

The twin villages of Codford St Peter and Codford St Mary, now referred to jointly as 'Codford', would also witness a dramatic change. By the end of 1914, the 26th Division was using tented camps in the area at the Fisherton Delamere end of the villages. Also affected were the villages of Stockton, Sherrington, Upton Lovell, Boyton and Corton. In September 1914, some 24,000 troops of both the 25th and 26th Divisions exercised and camped in the area of Codford, mainly using Manor Farm as their base, with the 2nd Wessex Royal Engineers camped near Codford station.

In due course, the tented camps at Codford were replaced by 15 hutted ones, named Camp No. 1 to Camp No. 15, all of which, when eventually finished, were originally occupied by the 25th Division. Along the Wylye Valley, a further 12 camps were built around the villages of Sherrington, Boyton, Upton Lovell and Corton, all of which were allocated to the 26th Division. Another camp which subsequently became permanent was at Knook, occupying an area of 50 acres and used as an artillery camp. Heytesbury House, the home of Siegfried Sassoon, was taken over for officers' quarters and Codford Camp Railway was laid. Straddling the River Wylye, the

25 *Camp No. 5 at Codford St Mary, with soldiers practising signals.*

26 *Imber Court, the home of Tom Dean, one of many properties used during the war to house troops. Forty-six foot soldiers, including six batmen, were housed in the attic, while a major and five other officers enjoyed the more stately accommodation offered by the mansion itself. Every morning before the soldiers departed, at 6a.m., Gladys Dean and a maid would provide tea for the men.*

camps at Codford, especially those close to the station were prone to flooding, and soldiers who stayed there afforded it the nickname 'Codford-on-the-Mud'. The 26th Division was eventually posted to Salonika, while the 25th Division went to France.

Evidence suggests that, as well as official camps, the War Office had the authority to billet soldiers wherever they saw fit. At the heart of the remote village of Imber, for instance, was the local country estate, Imber Court, providing much of the community's employment. Imber Court was the home of Tom Dean, away in the Army himself, who left his sister Gladys to manage the estate. One day an Army officer arrived to inform her that troops would have to be billeted in her home. Forty-six foot soldiers, including six batmen, were to be housed in the attic, while a major and five other officers would enjoy the more stately accommodation offered by the mansion itself. Every morning before the soldiers departed at 6a.m., Gladys Dean and one of her maids would provide tea for them, and concerts intended to improve the morale of these men would be given in the great barn, at which Gladys Dean sang and played the piano. She would also provide the accompaniment as the audience took turns at singing the patriotic songs of the time.

No doubt inspired by Gladys Dean, other Imber folk opened their doors to soldiers camping on the downs, providing a welcome relief to lonely men a long way from home. This was highly commendable, considering that the lives of the villagers had been turned upside down. They were only allowed to leave the village three times a week to go shopping in Warminster and shellfire is said to have damaged some of their cottages, as well as the track to Heytesbury. Most of this damage probably occurred after the adjacent Chapperton Down was used as an artillery school. The vicar at the time, Charles Watling, described Imber as 'a shell shocked village' dependent upon neighbouring villages for their food supply.

There were never enough dry billets for all the soldiers accumulating in the area, and thousands still had to suffer the cold, miserable existence of living in tents in one of England's bleakest environments. In his book *The Time Has Come: Officer and Temporary Gentleman*, Dennis Wheatley likened living in a bell-tent on Salisbury Plain to 'sheer purgatory'. Wheatley was serving with 2/1st City of London Royal Field Artillery, which was camped to the south of Imber in November 1916. He wrote:

> The rain increased until it was continuous. Day after day it poured in torrents while the men, protected only by their mackintosh ground-sheets, laced around their necks, toiled at digging gunpits. Our chaps had never before been called on to live in the open, so our cooks had no experience of using camp ovens. Petrol and sugar was flung on the fires in vain, time after time wind-driven rain put them out. During lulls they managed to boil up kettles for tea but nearly all our food had to be eaten cold. On that desolate plain there was not a house or barn in which anyone could shelter even for a short while. Drenched to the skin, their boots sodden, unwashed and utterly miserable, the men crouched shivering in their bivvies.

Conditions were so appalling for some of these troops that many suffered bronchitis, pneumonia, even death. Two soldiers from the 2/1st City of London Royal Field Artillery died and, from a total force of 498, 16 officers and 170 men had to be evacuated because of uncontrollable coughing. After eventually moving to France and taking part in several battles, Wheatley himself developed bronchitis in 1918, aggravated by chlorine gas, and was invalided back to hospital in England.

Another killer disease to hit soldiers stationed on Salisbury Plain was meningitis. In his book *Last Post*, historian Max Arthur recorded the memories of First World War veterans. One of them was Fred Lloyd, from Copwood in Uckfield, who joined the Royal Field Artillery in February 1916, when he was just 18 years old. Although he and his mates were young, fit and healthy, they were powerless against the effects of the crippling disease. Fred Lloyd wrote:

27 *General view of the camps at Sutton Veny.*

28 *A pay parade supposedly at Heytesbury during the First World War, most probably in the area where Knook Camp stands today.*

> We did three months training on Salisbury Plain at Stonehenge. The riding master was an old sweat and he just shoved me on the horse and said, 'On you go.' I started off very well but there was a little jump in front of me, and when we got there, he just flicked his hunting crop on the horse's tail and it shot me off. So that was a good start. He made me get back on – you had to be made of tough stuff. The horses were big – good horses. They were hunters mostly – they must have had eight or nine.
>
> It was hard, but I was a fit, strong lad, and used to heavy outdoor work, so I could take it. Most of the other lads came from the slums of London and were in poor shape – ill and undernourished. At the end of the Field Artillery training there was an outbreak of meningitis. It was dreadful. We were all struck down by it and only two of us survived. The rest died or were discharged as unfit for Army service because of the shocking after-effects. Many were blind and nearly all of them had lost their hearing – stone deaf for the rest of their lives. I spent months in hospital and was in very poor health for many a year. I'd been A1 fit before, but after I recovered I was classed as C3 – not fit for front-line duty – so I was transferred to the Veterinary Corps. That may have saved my life, because I was deemed unfit for the trenches. Even so, it was no picnic – in fact it was hell on earth.

Continued Growth

By 1914 Tidworth was beginning to rival Aldershot in importance, particularly in view of the fact that it sat within the confines of the country's finest training area. Tedworth House, which had been the grand residence of the commander-in-chief of Southern Command, became an officers' club in 1915. During the First World War, to take advantage of the influx of troops, a parade of shops was built along Station Road which became known to soldiers as 'Robbery Row'. A bank was also built, the architect of which was Sir Edward Maufe, who went on to design Guildford Cathedral. Also around this time, the light military railway was extended and ran across the main road to the barracks themselves.

Before the war several welfare institutes had been established in the town, and in October 1914 they were joined by the Red Shield Services Club in Baroda Road, which was administered by the Salvation Army. Also in 1914, the YMCA built several wooden huts where soldiers could relax when off duty. Examples of these could be found in Lucknow, Candahar, Aliwal and Bhurtpore Barracks, and a 612-seat Hippodrome Cinema was built somewhere around this time in Pennings Road, which unfortunately burned down in 1963.

But, once again, the area around Larkhill would enjoy the greatest improvements in terms of leisure facilities for the men. The YMCA provided five huts distributed around the 34 camps, with others at Rollestone, The Bustard and Tilshead. The Salvation Army provided two huts, one at Larkhill Camp, known as 'Larkhill I', and the other at Durrington known as 'Larkhill II', and in 1915 the military cinema was built on the north side of The Packway, which in 1919 was also destroyed by fire but later replaced by a garrison theatre.

The very first Church of England garrison church at Larkhill was also built in 1914, to the east of Lawrence Road. This timber and corrugated iron structure was erected at the same time as the hutted camps, and it's quite likely that the first Roman Catholic church was erected at the same time.

At Bulford the YMCA provided four huts, and the Salvation Army a further four. The first of these, known as 'Bulford I Salvation Army Hut', was erected

29 *This picture was taken on the Chitterne Road, which runs between the two parts of Knook Camp. It was posted in July 1915 and shows an artillery regiment.*

30 *Anzac troops relaxing at the Church of England Institute in Heytesbury.*

in B Lines in 1914, the second, 'Bulford II', was opened in 1915 near the Miss Perks' Home, and the third was 'Sling Salvation Army Hut', opened at Sling Camp in the same year. The fourth hut, opened in March 1918, is of particular significance as it was the very first Salvation Army hut specifically created for the Women's Auxiliary Army Corps. This was built behind 'Bulford I Hut' and would have been somewhere in the area between Haig Road and Wing Road in modern-day Bulford Camp.

The camps at Sutton Veny were serviced by a YMCA establishment at Greenhill House and three separate cinemas, one in Five Ash Lane, another in a local quarry, and the third in Dymocks Lane. The quarry cinema was burned to the ground during a riot, when unhappy soldiers were told by the proprietor's wife one evening that they couldn't have a refund after the projector had broken. The lady in question had previously told her husband to make himself scarce with their money, and all she offered was a free ticket for a future showing, knowing full well that most of the men involved were about to sail for France. Despite the attempts of a young officer to intervene, the cinema was sacked and burned. The piano used to accompany the silent films was dragged outside, and after the men had entertained themselves by singing songs, including 'Keep the Home Fires Burning', this too added fuel to the flames.

It's hard to know exactly what any of the YMCA establishments were like, as many of them were merely converted huts, but from publicity produced at the time for Greenhill House, called Sutton Veny House today, it's evident that the Australian soldiers relaxing here could expect something a little more spacious:

> Situated in the midst of the largest Australian International Force training and convalescent depots in Great Britain, this mansion, containing over 500 rooms with spacious lawns and grounds, offers ideal possibilities for rest and recuperation for war weary troops. The Home and grounds are now in the possession of the Australian YMCA and present a homelike aspect to officers, NCOs and men respectively.

In Codford, the YMCA erected a marquee that could accommodate a thousand relaxing soldiers, and after six months and thanks to an anonymous donor it was made into a permanent structure. Similarly, the church at Boyton opened its doors as a place where men could relax, and at Heytesbury there was a Church of England Institute in a wooden hut. But, of all these villages, the appearance of Codford must have changed the most, as temporary shops lined the streets, even being erected in the front gardens of cottages. All were aimed at separating soldiers from their hard-earned cash while, at Sutton Veny, snack shops and tea huts were built for the mutual good. Local women at many villages earned a little extra money by taking in soldiers' laundry.

Because of the number of soldiers now living in the area, and the high incidence of things like bronchitis and pneumonia, extra hospitals were also required. In 1914, at the northern end of Fargo Plantation between Larkhill and Rollestone Camp, was the 1,200 bed Fargo Military Hospital, which also included accommodation for the nursing staff employed there. At Sutton Veny a hutted hospital opened in 1916, containing beds for 11 officers and 1,261 soldiers. A Red Cross Hospital was established near Codford St Mary church, followed by Codford Military Hospital situated in Punch Bowl Bottom. And, in 1917, the 950-bed Australian Forces Hospital was built in Horne Road, Bulford, where a

31 *A British soldier is seen here riding through the quiet village of Sutton Veny.*

32 *A column of soldiers marching along Amesbury's High Street.*

number of huts were erected on the adjoining site between what is now Kashmir Road and New Ward Road, presumably for the hospital staff. This area was subsequently named Somme Lines.

Training for War

The wartime army on Salisbury Plain continued to grow, as battalion after battalion arrived at the new hutted camps, or came to live under canvas on the downs. Here their training would begin in earnest, with frequent rifle practice, route marches and trench digging. They would also take part in field days and week-long exercises.

In the summer of 1915, the 26th Division underwent a period of intensive training. Every day regiments would march behind their bands onto the Plain, complete with field kitchens drawn by mules. The kitchens had a fire-grate and chimney and, as the column marched along, the food would be cooking gently away. Fields near Sutton Veny were used to practise trench digging and trench warfare, with men living in trenches for a week at a time, simulating wartime conditions. All supplies, including food and ammunition, were brought up to them, as they would have been in the field.

Part of the division was made up of the 78th Infantry Brigade, which consisted of the 9th Battalion The Gloucestershire Regiment, 11th Battalion The Worcestershire Regiment, 7th Battalion The Ox and Bucks Light Infantry, and 7th Battalion The Royal Berkshire Regiment. The brigade gathered at Fovant before moving to Sandhill Camp at Longbridge Deverill on 21 July 1915 to be near the other units of the division. Many of the soldiers didn't even have uniforms, which slowly began to reach them during the course of their training. The 7th Battalion The Royal Berkshire Regiment didn't receive khaki uniforms until they were camped at Fovant, and wore blue post office workers uniforms from the time of their formation at Reading in August 1914, as these were the only garments available to them. Suitably attired, the 26th Division was inspected by King George V at Warminster on 25 June 1915, sailing from Southampton on 15 September for Boulogne, shortly to be replaced by the 34th Northumbrian Division. The 26th only stayed in France for a short period, as in November they were posted to Salonika.

Rifle ranges were built at Durrington and Rollestone, while a School of Musketry opened at Larkhill. At Rollestone there were four 600-yard ranges, while Durrington boasted five, sited between 700 yards and 1,000 yards. Although there were several prewar ranges near Larkhill, at locations such as Shrewton Folly, Wilsford Down, Rushall Down and Honey Down Barn, by this time these had already become redundant.

In *Last Post* by Max Arthur, First World War veteran Harry Patch, who served with the Duke of Cornwall's Light Infantry, recalls a period of training at Sutton Veny with a training battalion. He remembers practising on a 600-yard rifle range, although it's possible that the troops may have been taken to one of the ranges previously noted rather than one at Sutton Veny itself. He wrote:

> When the war broke out I was sixteen, and at eighteen I was called up. I had six months training at Sutton Veny, just outside Warminster, with the 33rd Training Battalion. On the rifle range, I needed eight points to be a marksman. I may not have been interested in being a marksman, but I was interested in the 6d a day extra. Anyway, I needed eight marks, and I said to the sergeant, 'This rifle is firing high, can I lower the sights?' We were on the 600 mark range. He said, 'I don't know. I'll ask the officer.' So he came and he said, 'Lower them to 550.' I did and my next two shots were bull's eyes. That gave me twelve points – four more than I needed so I didn't trouble a damn where my next shots went – I think I missed the target. Anyway, when I got to France, I was put on the Lewis Gun.

In 1916 the Overseas Artillery School was located on Chapperton Down, an area which lay between Tilshead and the village of Imber. This range would later be called the Chapperton Down Artillery School, part of the present Imber Ranges. The requirement of the school was to standardise the terminology used by its multi-national students. The school's headquarters were situated in Fisherton Street, Salisbury, and as there was no accommodation on the downs for both students and staff, they had to be billeted elsewhere, including the *Old*

George Inn and *The White Horse Hotel* in Salisbury, from where they were motored to the range each day.

Throughout the First World War, Larkhill remained one of Britain's most important training centres for the Royal Artillery. Standard training – on the gun park, ranges, and in the classroom – of newly joined gunners was maintained, but Larkhill also became a refresher centre for those already serving in Europe. In March 1918, for instance, the men of No. 96 Brigade, Royal Field Artillery, were recuperating at Larkhill from service in France, with most of their complement either wounded or classed as low category when, suddenly, they were ordered to mobilise in preparation for the great German offensive that was expected. There was a period of intense firing, remounts were arranged, and the men were medically examined. Although many were wounded and recuperating, they chose to be listed A1 for overseas service, and on 19 March they entrained from Amesbury station, arriving in France to serve with the 5th Army. The German offensive was subsequently halted and the brigade went forward again on 8 August, serving with General Plumer's 2nd Army from then until the Armistice on 11 November.

33 *Learning to dig trenches in fields near Sutton Veny.*

Throughout the war, Salisbury Plain remained one of the principal areas for experimenting with aerial warfare, and in December 1914 it was decided to form an Experimental Flight at the Central Flying School at Upavon. In charge of this experimental unit was Captain A.H.L. Soames, formerly of The King's Own Hussars, who was prematurely killed in June 1915 when a 500lb bomb on which he was working blew up in his face. Because of this, armament testing was stopped at Upavon and transferred to Orfordness, although the flight continued with other aspects of experimentation. Soames' second in command was Lieutenant G.M.B. Dobson of the Royal Flying Corps, who before the war had worked at Upavon as a civilian meteorologist. Other notables who worked there included

34 *Throughout the war, Larkhill remained one of the country's most important areas for training the Royal Artillery. This image is from a postcard, sent home in September 1916 from a soldier in the RA.*

an Oxford don, Lieutenant R. Bourdillon, Lieutenant H.T. Tizzard and Lieutenant F.A. Lindemann (later Lord Cherwell), who was Winston Churchill's personal assistant during the Second World War.

In October 1916 the Experimental Flight was renamed the Testing Squadron RFC, and experiments of many types were carried out. One of their earliest achievements was the production of a bomb sight, later used with great effect on the Western Front. In January 1917 the Testing Squadron was transferred to Martlesham Heath near Ipswich, which became the Aeroplane Experimental Station, but this didn't stop experiments taking place on Salisbury Plain. In March 1917, the first attempt at using a remotely controlled bomber took place at Upavon, unsuccessfully, and on 21 March 1917 the first British guided missile anti-tank weapon designed by Professor A.M. Low began flight trials at Upavon.

The vulnerability and exposure of Salisbury Plain's airfields was most apparent during storms, the worst of which was the Great Gale that occurred between 4 and 5 November 1918. On that particular occasion, worst hit was Upavon, where no fewer than 17 aircraft and two hangars were destroyed. At some point during the storm, two Handley Page aircraft actually took off under their own steam, landing again about a half a mile away!

As more and more Royal Flying Corps squadrons were formed there was an increasing need to train pilots, and the Central Flying School soon became unable to cope. To support the demand and train officers for their aviator's certificate, Netheravon was temporarily chosen as an extension of the Central Flying School. But the responsibility only lasted until January 1915, when trainee pilots were transferred to Shoreham. By now, Netheravon was required to be the formation base for squadrons under the command of No. 4 Wing, who were getting ready for overseas service.

35 *Uniform of a corporal in the Royal Horse Artillery during the First World War. From a painting by Alix Baker PCAFAS FRSA.*

During the remainder of the war, several new airfields were opened on Salisbury Plain, although one of these, situated some 800 yards east of the Tilshead-West Lavington road, was only a temporary site used by No. 105 Squadron RFC before it departed for France in December 1915. Of the others, Boscombe Down, originally known as Red House Farm, would eventually emerge as the most significant, although when it first opened it was only for a short period of time. It first opened on 1 October 1917 as No. 6 Training Depot Station, where Royal Flying Corps personnel were trained in day bombing procedures. The airfield, which is situated to the south-east of Amesbury, had a railway connection, but its hangars weren't finished until November 1918. After the war ended, the site was closed down in April 1920 and returned to farming, having converted from an operational aerodrome to a storage unit in May 1919.

Probably the second airfield in importance was Old Sarum, two miles north of Salisbury, which like Boscombe Down was used to train personnel in day bombing procedures when it opened in August 1917, and was later designated No. 11 Training Depot Station. Lake Down airfield, also opened in August 1917, was located two miles south-west of Stonehenge, opposite Druid's Lodge. It was the very last stop on the Larkhill Military Railway until its closure. As No. 2 Training Depot Station, it, too, was used for training both British and, later, American pilots in day bombing. The station had a short life as it closed in October 1919 with the majority of the buildings being demolished, although three structures have survived on the side of the A360. Photographs of the time show that it had a large technical site and rows of white bell-tents for squadron accommodation.

The aerodrome at Stonehenge itself, which was roughly a quarter mile to the west of the ancient monument, was used by the School of Navigation and Bomb Dropping. Some of the hangars here were called the Handley Page hangars and were used by the Royal Naval Air Service, who flew Handley Page aircraft. The station closed in 1920 and in time all evidence of its existence disappeared in order to protect the visual integrity of Stonehenge itself. There was a further Royal Flying Corps Training Depot Station at Lopcombe Corner along the A30 between Salisbury and Middle Wallop. It was here that No. 74 Squadron, Royal Air Force, disbanded after returning from France in 1919, where they earned the nickname 'Tiger Squadron'. On 1 April 1918 the Royal Flying Corps and the Royal Naval Air Service had amalgamated to become the Royal Air Force.

Finally, there was an airfield at Rollestone which was used for training with balloons and kites as, although the aeroplane had taken over from these for artillery observation purposes during the course of the war, they were still being used to a certain degree. The Royal Flying Corps opened a balloon school in July 1916 at Rollestone Bake Farm near Larkhill, No. 1 Balloon School, which had an outstation at Tilshead. The balloons were attached to the ground by guy ropes and their baskets were provided with parachutes, just in case the occupant had to make an emergency descent.

36 *The King, with his staff and standard-bearer, passing the ruins of Stonehenge during an inspection of the troops on Salisbury Plain.*

Soldiers from Around the World

Citizens of the Empire and Dominions also answered the call and were soon arriving on Salisbury Plain. Four camps on West Down became occupied by troops from Canada and Newfoundland from October 1914. The following January, a Canadian Training Depot was established at Tidworth, and the Cavalry School was closed at Netheravon House and used as a convalescence home for Canadian troops. Although Newfoundland is a Canadian province today, until 1949 it was an independent British colony. The first Newfoundland contingent of 500 men sailed to Britain with the original Canadian force, and was later enlarged into the Royal Newfoundland Regiment.

Canadian troops had helped to build Sling Camp, Bulford, and, in fact, the Canadian Expeditionary Force was the first large group of Dominion troops to arrive in the area, having landed at Plymouth and disembarked between 15 and 24 October 1914. This force, the 1st Canadian Division, subsequently proceeded to the four tented camps on West Down, but because their arrival coincided with a period of incessant rain it was decided to move some of them to the partially built camp at Sling.

On first arriving on the Plain, the four camps occupied by this Canadian and Newfoundland force were Bustard Camp, West Down South Camp, West Down North Camp and Pond Farm Camp. The division consisted of the following: 1st Canadian Infantry Brigade, the divisional mounted troops and Princess Patricia's Canadian Light Infantry, who occupied Bustard Camp; the 2nd and 3rd Infantry Brigades, who took over West Down South Camp; the divisional artillery and divisional supply column, who took over West Down North Camp; and the 4th Infantry Brigade, the cavalry, the Newfoundland contingent and

37 *Captain in the Royal Flying Corps, c.1917, with a Sopwith Pup aeroplane. From a painting by Alix Baker PCAFAS FRSA.*

any other miscellaneous troops occupying Pond Farm Camp. The divisional artillery consisted of three field brigades each deploying three six-gun batteries, armed with 18-pounder guns.

The Canadian cavalry of this expeditionary force were the 19th Alberta Dragoons, the Royal Canadian Dragoons, Lord Strathcona's Horse and the British 2nd King Edward's Horse (formed from colonials in Britain). There was also the Canadian Light Horse, consisting of four militia squadrons, one of which was the Royal North-West Mounted Police. There was also the 1st Automobile Machine-Gun Brigade, which consisted of armoured cars mounted with Colt guns, and 13 mounted infantry regiments as follows: 1st Brandon, 2nd Victoria, 3rd Medicine Hat, 4th Toronto, 5th Sherbrooke, 6th Amherst, 7th London, 8th Ottawa, 9th Lloydminster, 10th Portage, 11th Vancouver, 12th Calgary and 13th Pincer Creek. All these troops would train on the Wiltshire downs towards the end of 1914.

Because of the torrential rain, the four tented camps soon turned into quagmires and, with the tents being largely unheated, the soldiers were soon soaked and freezing cold. The mud was ankle deep and there was no method of drying clothes. With people going sick in large numbers, the situation became critical. With the departure of the New Zealanders working at Sling the work was being continued by civilian contractors, but with the aid of skilled craftsmen from the Canadian Force it was soon in hand. Other Canadian soldiers with building experience were also lent to the construction of camps at Larkhill.

So, after only a short tenancy at their tented camp at Pond Farm, the 4th Infantry Brigade was able to move to the more hospitable huts at Sling Camp on 9 November 1914. But this still left most of the force under canvas. The problem was alleviated slightly when Princess Patricia's Light Infantry proceeded to France with the 27th British Division, after which their tents at Bustard Camp were taken over by the Newfoundland contingent. After a very short stay themselves, the Newfoundlanders marched to Amesbury station on 7 December, where they entrained for Fort George, Inverness. What weather conditions greeted them there we can only guess at.

When towards the end of December huts became available at Larkhill, these were used by the 2nd and 3rd Infantry Brigades and the Canadian Engineers. But this still left roughly 11,000 troops under canvas on Salisbury Plain, in atrocious weather conditions. The final solution was to billet the remaining troops with local communities, a scheme that covered a considerable area. The

Royal Canadian Horse Artillery was billeted around Bratton and Erlestoke, Lord Strathcona's Horse around Upavon and Pewsey, the Royal Canadian Dragoons between Tilshead and Wilton, and the divisional artillery around the Market Lavington, Devizes and Rushall areas. This left mainly the 1st Canadian Infantry Brigade under canvas by January 1915. All this seems a little too late, as in February 1915 the Canadian Expeditionary Force left for France via Avonmouth. However, shortly before their departure they were reviewed on Salisbury Plain on 4 February 1915 by the King and Lord Kitchener.

When the Australian Imperial Force (AIF) entered the war, they went straight into the Gallipoli campaign after conditioning in Egypt, following which they found themselves destined for France. At this point, training battalions were established in England from where the troops on the front line could be reinforced. Three separate reinforcement battalions were created on Salisbury Plain, to supply trained soldiers to complement four divisions in the field. Soldiers destined for the 1st Division were trained at Perham Down, those for the 2nd and 3rd Divisions were trained at Rollestone Camp, while recruits for the 5th Division trained at Larkhill. There was also an Australian Artillery Deport at Larkhill, a Machine Gun Corps Depot at Tidworth Barracks, and depots for engineers, signallers, Army Service Corps personnel and the Army Medical Corps at Parkhouse Camp.

Recruits began to arrive at the training battalions in June 1916, and by the following August Bhurtpore Barracks had become the headquarters of the Australian Imperial Force Depots in the United Kingdom. Before going to France in July 1916, the entire 3rd Australian Division was briefly stationed at Larkhill and one of four command depots was established at Perham Down to look after men who had been discharged from hospital processing soldiers considered fit to return to front-line duties. The Australian Reserve Brigade was stationed at Knook Camp. After the war the Australians placed flags in Heytesbury church and members of the brigade presented a lectern bible to the church before departing.

No. 1 Australian Command Depot moved to Sutton Veny in December 1916, with the Overseas Training Brigade of the Australian Imperial Force camped at Longbridge Deverill. This transferred many more

38 *Ashley John Gillman (1885-1969), who was born in New South Wales, Australia, the son of William Gillman and grandson of John Gillman of Foss Farm, Brokenborough. Ashley came to England with the Australian Imperial Forces in 1917 and was stationed at Larkhill before moving to France. After being wounded he was sent to Orpington in Kent. Evelyn Dancey explains that Ashley was her mother's second cousin and his own son Ian Gillman has been to visit her family several times from Brisbane.*

39 *In this photograph an Australian Army band is seen marching past the* Angel Inn, *Heytesbury, sometime during the First World War.*

Australians into the hutted camps that had been built on Sutton Common. It also put huge pressure on the military hospital at Sutton Veny, now No. 1 Australian General Hospital, when a particularly virulent strain of influenza struck the camps after the 1918 Armistice. The fighting was over but this strain of influenza, known as Spanish Flu, was claiming lives at an unprecedented rate. Part of the churchyard at the parish church of St John the Evangelist was made into a War Graves Cemetery, and a memorial service is held there each year on the Sunday that falls closest to 25 April, or Anzac Day. In total there are 167 people buried in the war graves section of the churchyard, and inside the church is a memorial chapel listing the names of 142 Australian soldiers who perished here, along with two nursing sisters, one of whom, Matron Jean Miles Walker, was also Australian. These deaths were not all caused by Spanish Flu, but certainly a large proportion would have been.

The graves of other Australian soldiers who died on Salisbury Plain and the surrounding area can be found in other cemeteries, including Fovant where there are at least thirty-nine. In the military cemetery at Tidworth below Sidbury Hill are the graves of 150 Australian and 100 New Zealand soldiers who died at either Tidworth Military Hospital or Fargo Hospital. And there is another Commonwealth War Graves Cemetery at Codford.

40 *The map of Australia carved on a hill at Compton Chamberlayne.*

Australian soldiers carved a large map of their island country in the chalk at Compton Chamberlayne, which for many years had a flagpole at its centre from which the Australian flag was proudly displayed. And a reproduction of the badge of the Australian Forces was cut on the side of Lamb Down hill above Foxhole Bottom, alongside the A36, which can still be viewed today. Although No. 1 Australian Command Depot was at Sutton Veny, some troops were based at Codford, and the site at Lamb Down was probably chosen as it had a more prominent position. A board at the site also suggests that the commander of the Australian brigade chose the hill because he could see it from the reading room at Stockton House where he was staying. Some suggest he wanted to leave his country's mark on the area, while others claim the badge was carved by way of punishment for soldiers based with the 13th Training Battalion who would be made to climb the hill and cut the chalk wearing full packs and equipment. Because of this, the spur on which the badge is carved became known as Misery Hill. When it was finished, the image was adorned with green, brown and clear beer bottles, to make it shine in the sun like bronze, the colour of the actual Australian cap badge. In time, the glass was either washed away, or sank into the chalk below.

After the war, Salisbury Plain became a mustering ground for around 80,000 Australian soldiers waiting to be repatriated. The last contingents of men arrived from France in May 1919, and stayed mostly at Bulford and Tidworth

41 *Larkhill in the mud during the First World War. Terry Crawford notes, 'Some soldiers bought rubber waders at exorbitant prices from dealers. The Canadians improvised snow-ploughs to push the mud away from some roads on the Plain but with limited success. The sender of this card wrote, 'What do you think of the mud it couldn't be any worse in France … it is worst than this at times we have to go about like ducks.'*

until arrangements were made for their return home. At Sutton Veny, the Australians formed an agricultural training depot in order to prepare and train their soldiers for return to civilian life. By mid-1919, 2,000 soldiers had passed through it.

Soldiers from New Zealand also arrived, who like the Australian forces had been fighting in Gallipoli. A training depot was established at Sling Camp where they left behind a permanent memento in the Kiwi carved on Beacon Hill. New Zealand casualties from the fighting against the Turks were initially brought to a convalescent centre in Hornchurch and, once fit again, were sent to an Australian depot in Weymouth. By March 1916, there were already around 2,000 men convalescing in Britain. Hornchurch was the only New Zealand camp in Britain, and had to accommodate both fit and unwell soldiers. The decision was then taken to keep the convalescing soldiers at Hornchurch and send the recovering men, those who were likely to return to service, to a camp at Codford. Those who were well enough for posting were sent to the training camp at the 4th New Zealand Infantry Brigade Reserve Camp at Sling, which not only retrained battle casualties, but reinforced service battalions with newly trained recruits. Sling continued as the main New Zealand training camp throughout the rest of the war and, although it was hutted, it was bleak and unpopular. So, in order to give it a cheerier feel, the YMCA established a hut to entertain the men, which included a cinema.

Codford became the New Zealand Command Depot in July 1916, where wounded soldiers were sent to convalesce, and the military hospital at the village became No. 3 New Zealand General Hospital, with beds for 10 officers and 980 soldiers. At this time Codford also became the scene of the New Zealand War Contingent's Club, which was one of the most popular and best-equipped welfare clubs on Salisbury Plain and was run by ladies of the War Contingent Association. In October 1918, there were roughly three and a half thousand soldiers at the depot and hospital waiting repatriation. There are the graves of 66 New Zealanders and 31 Australians in the War Graves Commission Cemetery at Codford.

New Zealand forces could be found at many locations on or around the Plain, the most unusual of which were the Maori battalion who arrived at Southampton from France in the first week of January 1919 and, after travelling by train to Amesbury,

42 *Commonwealth War Graves in St John the Evangelist churchyard, Sutton Veny.*

43 *The Australian cap badge carved into the hill at Lamb Down.*

marched to No. 5 Camp at Larkhill. The battalion remained there until 28 February when it left for Liverpool on its way home to New Zealand again.

The Kiwi on Beacon Hill was the brainchild of a certain Major L.W. Metcalfe, MC, who was sensitive to the fact that, after the Armistice, the 4,000 soldiers at Sling Camp were still being subjected to wartime military discipline as they waited for their troop ships home. This led to unrest amongst the troops, and, in order to curtail any future trouble, Metcalfe thought it would be a good idea for the men to cut a Kiwi emblem out of the chalk hill above the camp. No doubt he was inspired by other emblems and badges he had seen carved around the Plain. He approached his commanding officer, Lieutenant Colonel A.G. MacKenzie, who also thought it a good idea. With the subsequent approval of Brigadier General G.S. Richardson, the General Officer Commanding the New Zealand Expeditionary Force Establishments in the United Kingdom, and the War Office, an accurate drawing was undertaken by Sergeant Major P.C. Blenkarne who was a drawing instructor in the New Zealand Army Education Corps.

The drawing was plotted out on the side of Beacon Hill, and once completed the Kiwi was 420ft long, with a 50ft beak. Next to it were the letters 'NZ', which were themselves 65ft high. The whole thing covered about an acre and a half of land. The Kiwi is still there today, and since 1980 it has been looked after by 249 Signal Squadron (AMF (L)) of the Royal Signal Corps. This first came about because, after many years of neglect, a team under the command of Major J.H. Fisher, MBE, restored the lacklustre bird to its former glory. In August 1981 the New Zealand General Staff presented a carved model of a Kiwi to the squadron in recognition of their work, and the bird is still regularly maintained by them.

As for Americans, probably no more than a few hundred appeared around Salisbury Plain during the First World War. Terry Crawford, an authority on Wiltshire during this period, explains that:

> George Buchanan Fife, in *The Passing Legions: how the American Red Cross met the American Army in Great Britain, the gateway to France*, refers to the American Red Cross providing a canteen service at 'a large rest camp' at Codford and mentions an 'American construction camp' at Lopcombe Corner. It also provided facilities at Netheravon, Stonehenge, Upavon, Old Sarum, Lake Down and Yatesbury airfields, and camp infirmaries at the last three.

On pages 111-22, Fife explains how the main work of the Red Cross on the 'Salisbury Plain airfields' was to set up camp dispensaries (shops rather than medical facilities) in marquee tents, two to a post; these were to be replaced by portable wooden 25 by 50ft huts for winter use, but these had been installed in only two posts by the Armistice. The American Red Cross then devoted much time to providing recreation and amusement, supplying uniforms and equipment for baseball and arranging inter-camp matches. It also stocked 81 different articles, from sweaters to razor blades. The US YMCA also provided facilities at airfields where Americans were based and had a store at Chaplin's, in Salisbury.

References to American soldiers in Wiltshire during the Great War are few, though US pilots trained at airfields in the county. Edwin A. Pratt, in *British Railways and the Great War*, states that on arrival in Britain some members of the American Expeditionary Force went to Bulford and Fovant, but the context suggests that the numbers were small. The Order of Battle volume listing 'general headquarters – services of supply and separate forces' mentions an 'air service camp', part of Services of Supply Base Section No. 3, established at Codford on 23 September 1918; one or two of its staff were involved in a fracas with New Zealand troops at the *George Hotel* in Codford in November 1918.

After the Armistice

After the Armistice, although the war was technically still ongoing, different locations dealt with the new situation in different ways. As already seen at Sling Camp, New Zealand soldiers were still subject to wartime training and discipline, while the Australians at Sutton Veny were winding down and preparing to go home. The British military presence in the Warminster area reduced rapidly to a troop of the Royal Wiltshire Yeomanry and a detachment of the 4th Battalion of the Wiltshire Regiment.

Around the Imber area, mules became a familiar sight as various armies sold them off to the locals in large quantities. Although they were ideal used as transport, pulling such things as domestic traps, agricultural carts and water barrels, they had a tendency to escape and eat anything in their path. In his book *Little Imber on the Down*, Rex Sawyer notes that a gentleman called Cyril Nash, the son of the village blacksmith, had been casually employed by the military authorities during the war because of his local knowledge. His duties included

44 *12th Battalion the Argyll and Sutherland Highlanders at Sutton Veny, First World War.*

locating water, digging drains, and looking for officials who had strayed too far on the downs and become lost. While executing these duties he had often come across army mules, and vowed to the drinkers in the *Bell Inn* that he would never own one for, although they could stand double the work of any horse their size, they 'kicked in every direction like a cow'.

At the time of the Armistice, there was a new vicar in the parish, Edgar Glanfield, who recorded the occasion in the churchwardens' book. A group of soldiers billeted in one of the local farms approached him to ask whether they could ring a peal of bells as a thanksgiving? The 13th-century church of St Giles had magnificent bells, cast by John Lot of Warminster. Subsequently, three of the bells rang all day long, the others being too badly cracked for use, and Glanfield held a service for the soldiers in the evening.

Rex Sawyer also records how, on Armistice Day, the local people of Salisbury Plain became just as caught up in the euphoria of the moment as the soldiers billeted there:

Mrs Mullings of Edington took the bread to Imber on Armistice Day as she had many times before during those difficult years. With Mr Marsh, the baker, they drove two bread carts, he in front with the lighter wagon, she following behind with a heavier covered van. From Bratton to Imber, up the steepest part of the escarpment, an extra horse was used for the van. Imber's celebrated home-made wines, parsnip, dandelion and many other varieties, were liberally offered along the way – 'very strong they were and very good'. All day long they were fêted. At Imber Court they joined a particularly jolly party. Australian soldiers, unaware of the fatal Spanish Flu soon to reduce drastically their numbers, were celebrating their anticipation of a swift return home.

It was nine in the evening before they started their woozy return, Mr Marsh with his chestnut-coloured nag, Jimmy, once more leading the way, and Mrs Mullings with Jacky pulling the heavier van. By this late hour the track was difficult to manoeuvre and in the dark a piece of wool from the surrounding sheep blew into Jack's face. The horse took fright. 'I thought that was the end of me,' remembered Mrs Mullings. She put on the brake as the horse started to bolt. Fortunately he pulled up on reaching the cart in front and they returned home safely.

45 *This photograph shows men of the Royal Field Artillery during the First World War. The location is not known, but the scene looks as though it could be in one of the Wiltshire villages. At the very least, it illustrates the uniforms worn by these troops at the time.*

Chapter Three

Between the Wars

At the close of the First World War, life on Salisbury Plain began to return to normal and the victorious Army settled down to the serious business of peacetime soldiering. For the first few months, however, many of the wartime camps were used as transit areas for overseas troops waiting to go home. It was then a question of dismantling the camps, hut by hut, so that many disappeared altogether from the landscape, while others were retained for future use, some even being improved. All across the area, walkers can still see evidence of this remarkable period in our nation's history.

The inter-war years would see huge military changes, particularly in the Army, as mechanisation really came to the forefront of warfare and the era of the horse for all practical purposes drew to a close. Tanks and artillery would rapidly develop and Salisbury Plain was the ideal place to train with both. With this in mind, the War Office estate would continue to expand.

Soldiers at Peace

After the vast wartime Army had eventually gone from the area, Tidworth resumed its role as a garrison town. By the early 1920s, the 2nd Cavalry Brigade and the 7th Infantry Brigade were once again its main occupants, coming under the control of the General Officer Commanding the 3rd Division. This coincided with another period of change within the Army that saw numerous amalgamations taking place, particularly within the cavalry regiments, who were slowly becoming mechanised.

By January 1924 the 2nd Cavalry Brigade comprised the 12th Royal Lancers, the 14th/20th Hussars, the 15th/19th Hussars, the 17th/21st Lancers and the 2nd Signal Troop, Royal Corps of Signals. The 7th Infantry Brigade was made up of the 1st Battalion The Lancashire Fusiliers, 2nd Battalion The Loyal Regiment, 2nd Battalion The King's Shropshire Light Infantry and 1st Battalion The Wiltshire Regiment.

After the hard years of fighting in the trenches, soldiers now wanted to get back to peacetime soldiering and enjoy army life a little more. However, there were very few amenities in the camps, especially for sport. The garrison towns on Salisbury Plain had only been in existence for a few short years before the war erupted, and accommodation and military training facilities had been the priority with the War Office.

When the 17th/21st Lancers first arrived at Tidworth from Ireland in 1922, to occupy Aliwal Barracks, they were surprised by the lack of facilities. Other than one cinema, a few shops and very inadequate bus and train services, there were no proper facilities for sport and recreation. At Aliwal Barracks, all they found was one small, uneven football pitch which, surprisingly to them, was claimed to be the best in the garrison. The regiment decided this wasn't good enough and began to rectify the situation. With officers and troopers pulling together, they soon built two football pitches and a cricket pitch. Thousands of tons of chalk sub-soil had to be shifted, and turf was cut from Perham Down and brought to the site in limbered wagons. Inspired by this example, other regiments began to do the same and, once the individual sites had been finished, resident soldiers set out to build a number of sports grounds and arenas that would benefit the garrison as a whole rather than individual barracks. These would include the Oval, the cricket ground on the Shipton Road, and a proper polo arena.

Due to the fact that cavalry regiments played an important part in garrison life, Tidworth became noted for polo. In early years teams would have played each other in any suitable field, but the first ground reserved specifically for the game was set aside on the east of the A338, opposite St Mary's Church, in around 1918. Later, another area was provided immediately to the south of Tedworth House, and it was here that the first military tattoo was also held in 1920. Until 1925, both polo and subsequent tattoos took place here, until the

46 *The unveiling and dedication of the Cross of Sacrifice erected to the memory of the Australian dead at Sutton Veny church, 22 July 1923. One of the gentlemen in the foreground, is Sir Joseph Cook, High Commissioner for Australia. Terry Crawford notes that 'The Wiltshire section of The War Graves of the British Empire records 167 people buried in the war graves section (plus another British grave in the main part of the churchyard) 143 of which are Australian, including that of Matron Jean Miles Walker from Sydney. There was a 1,272 bed hospital at Sutton Veny Camp, and a significant number of the dead succumbed to the influenza of 1918-19.'*

47 *This postcard comes from the Tidworth Tattoo series printed in the late 1920s or early 1930s. This particular card is titled 'Bombardment of Toytown', and shows the lavish scale of these productions.*

officers and men of the brigade, working side-by-side, built a specially designed arena in Tidworth Park that could be used for tattoos and other recreational purposes, and incorporated an area for inter-regimental polo tournaments. The polo ground was christened the Fisher Ground after Brigadier B.D. Fisher, the commander of the 2nd Cavalry Brigade, who was instrumental in organising tattoos during his time in charge. On 8 July 1926, His Majesty King Alfonso of Spain visited the garrison to inspect 16th/5th Lancers, of whom he was colonel in chief. That afternoon he played polo for the regiment, helping them to beat the 7th Queen's Own Hussars.

The first Tidworth Tattoo took place in June 1920 in the form of a pageant, but it quickly became a very popular annual event. It was initially staged on the new polo ground near Tedworth House between 8 and 10 June 1920, and was largely the idea of Major General T.T. Pitman, CB, CMG, who commanded the 2nd Cavalry Brigade from 1920 to 1922. The scene was illuminated by searchlights placed on the roof of the house itself.

At the start, Pitman's ambition was to entertain troops from the garrison, their families, and any other interested local parties. There was no seating, except a few chairs provided for officers, but the event was a resounding success and raised £400 for military charities. It was so successful that further tattoos would be staged every year, until the outbreak of the Second World War. Although they would be revived at a later date, the last pre-war event was in 1938. Major General Pitman also organised the 1921 and 1922 tattoos, but in the following year the job was taken over by the commander of 7th Infantry Brigade, Colonel Commandant G.V. Hordern, CB, CMG.

A temporary grandstand was eventually built that could accommodate 200 people, but by the time Brigadier Fisher took over the running of the tattoos it was clear the event was out-growing the location provided for it. Money raised

48 *Detail from a postcard in the Tidworth Tattoo series titled 'Bombardment of Toytown'.*

from the 1924 and 1925 tattoos helped fund the building of some of the sports facilities in the garrison, and in 1926 the new arena was built in time for the July event, which could now accommodate 7,000 spectators.

When Fisher left Tidworth in 1926, the management of the tattoo was taken over by Southern Command, under the charge of Colonel W.J. Dugan, CB, CMG, DSO, and the event was renamed 'The Southern Command Tattoo'. As attendance continued to rise the facilities were constantly improved, until in 1929 a covered stand with seating for 1,025 spectators was built, incorporating a Royal Box, control platform, lavatories and storage facilities. This meant that around 13,750 people could now be seated, with additional standing room for another 10,000 spectators. This colourful military display continued until the war clouds of the late 1930s brought it to an untimely close.

As well as providing for sport, the 1920s and 1930s would see improvements in other areas of welfare. In Tidworth, the wooden huts built by the YMCA at Aliwal, Bhurtpore, Candahar and Lucknow Barracks were demolished and replaced by a permanent brick building at the junction of Bazaar Road and Dasna Road. Similarly, at Larkhill, the YMCA built a permanent brick building in 1935 on the north side of The Packway, between Watson Road and Whingates Road, which was named after Sir Arthur Yapp, KBE. This building, later given to the Army, is still there in the form of the Sharp Hall, named in memory of General Sir John Sharp who was commandant of the School of Artillery from 1964-6. Around this time the Salvation Army also established a Soldiers' Home on the north of The Packway, near its junction with Alanbrooke Road.

The Navy, Army and Air Force Institutes, or NAAFI, was formed on 1 January 1921. Its origins can be traced back to an organisation called the Canteen and Mess Co-operative Society, first formed by a group of serving officers in 1894 to provide canteen facilities for soldiers. During the First World War, this society helped form an Expeditionary Force Canteen Organisation, to provide facilities for soldiers in France. Later, the Royal Navy was included and the name changed to the Army and Navy Canteen Board; with the inclusion of the RAF, it became the NAAFI in 1921. Canteens would be established in all garrisons, and would be administered from an area office at 47 Endless Street, Salisbury.

At Bulford there was an annual sports day, when regiments would compete against each other. These were noted for being great family occasions, the wives and children of married soldiers coming to support their husband's team. The NAAFI would provide separate refreshment tents for warrant officers, sergeants and other ranks, while the officers' tent was particularly splendid, their salmon and cucumber sandwiches served on fine bone china, which the NAAFI called 'blue and gold'. As well as the obvious field sports, every year there would be a tug-of-war and a tent-pegging competition, when cavalry soldiers would hook up tent pegs from the ground with lances, while riding at speed.

At Tidworth and Bulford several canteens were run by the NAAFI, while at Larkhill the first NAAFI building was situated near the junction of Glover Road and Bell Road. This was later demolished and replaced by two establishments, one in Congreve Road and the other in Whingates Road. But the NAAFI also became known for providing tents or mobile facilities during summer camps for territorial troops or exercises on the downs.

49 *In this postcard in the Tidworth Tattoo series, it appears that the whole Army has been re-equipped for the enactment of a great battle from the past.*

50 *Assaye Barracks, Tidworth.*

Another organisation formed in the wake of the First World War was the British Legion. This association for ex-servicemen first came into being in 1921, and the branch in Tidworth was formed on 12 July 1922, with a building in Pennings Road shortly afterwards.

At Bulford, the Wesleyan Soldiers' Home was destroyed by fire in April 1922 and was replaced by the Cromwell Institute, which used a building purchased from the Congregational Church Authorities in December 1925. The Miss Perks' Home was taken over by a lady called Miss Elsie Sandes in 1926 and renamed the Sandes Soldiers' Home, one in a chain of institutes set up by this remarkable lady. She began her welfare work by teaching drummer boys to read at the Tralee Garrison in Ireland, but soon adult soldiers were using her home as a meeting place. The first Sandes Home was opened in 1869, when Elsie rented a house for soldiers to use for recreation, and by the time she died in 1934 they could be found all over Britain. She was buried in Tyrella Cemetery near Ballykinler Camp, with full military honours.

Further churches were also planned and built, Bulford eventually getting its Church of England garrison church in 1927. This was the church of St George, dedicated by the Bishop of Salisbury, the Rt Rev. St C.G.A. Donaldson on 3 November 1927. It was very much a Royal Artillery church, and quite appropriately the first turf had been cut by the commander-in-chief of the Royal Artillery, Brigadier General L.C.L. Oldfield, in June 1920. The foundation stone was laid in October of the same year by Field Marshal Lord Methuen, GCB, GCMG, GCVO. The pulpit was donated to the church in memory of Major General F.D.V. Wing by his widow, and the east window was given by the Royal Artillery Memorials Society in memory of all ranks of the regiment who fell in the First World War. A large round coloured window over the main door was paid for by public subscription, with local people, including children, giving varying amounts, but generally no more than a farthing.

Over the years there had been several Roman Catholic chapels at Bulford established in huts in different lines, but from December 1925 a corrugated iron building was used and dedicated to Our Lady of Victories. This was situated just to the north of the car park next to the present church of Our Lady Queen of Peace, which itself was opened to services in 1968.

At Larkhill, on the north side of The Packway opposite Lightfoot Road, was St Anthony's Roman Catholic Church, first used around 1931. This was later replaced by a hut on the north side of Gore Road at the junction with Alanbrooke Road, itself replaced by a later church in 1968. The Church of England garrison church of St Alban the Martyr was dedicated on 21 May 1938 by the Chaplain General the Rev. E.H. Thorold. The foundation stone had been laid by General Sir John Burnett-Stuart, KCB, KBE, CMG, DSO, in January 1937. This church, which is now very much regarded as the Royal Artillery's own church, was built on the north side of The Packway, between Lawrence Road and Glover Road.

Other developments within the garrison towns would reflect changes in the Army itself, as cavalry regiments slowly replaced their horses with armour. Despite the appearance of the tank on the Western Front in 1916, cavalry in the 1920s was still largely mounted, although by 1936 this situation had completely changed. In the 1930s, one eye-witness recalls, long columns of soldiers on exercise would march from the camp onto the Plain. The officers would be riding their horses, while the men marched. There would be mounted gun carriages and trains of mules, loaded down with supplies, being led by hand. As motor transport became more widely used, a motor ambulance might be positioned at the rear of the column, ready to evacuate any casualties to the hospital. But this sort of almost Victorian military procession was soon to disappear.

51 *Garrison church of St Alban the Martyr, Larkhill, in a view looking north-west across The Packway.*

Although the Cavalry School at Netheravon House had reopened in October 1919, it was short-lived, and in October 1922 the site was taken over by the Small Arms School (Netheravon Wing) which trained soldiers to use the Vickers Machine Gun, right up until 1939. The age of horses for practical military use was drawing to its conclusion. The 11th Hussars and the 12th Royal Lancers were the first cavalry regiments in 2nd Cavalry Brigade to lose their mounts, in 1928. From then on the brigade would be systematically mechanised. This necessitated alterations to many of the existing barracks, as stables had to be converted to either garages or parking areas for vehicles.

A certain amount of new building was also undertaken in the various garrisons, such as the creation of many more married quarters at Bulford. In 1937, 29 officers' quarters were erected at Sling, in Baghdad Road, Gaza Road and The Crescent, while 26 quarters for married soldiers were built in Hubert Hamilton Road and Marne Crescent. In 1938, 59 further married quarters were erected in New Ward Road and 30 in Melbourne Avenue.

For private soldiers and lance corporals, these red brick blocks of houses were numbered from A Block to E Block. Each house had a kitchen and sitting room downstairs, and two bedrooms upstairs. The bucket toilet would have been in a shed or lean-to at the back of the house. In the 1930s, outhouses were added to provide bathrooms. A road ran between the blocks, and at the front the houses had steel railings about three feet high and a gate that opened onto the road. F block was for corporals and acting sergeants, and H Block for sergeants and company sergeant majors. The regimental sergeant majors also had houses, usually in a block of four together. The larger detached houses for officers even had staff quarters, as sometimes a nursemaid was required. Officers also had a batman, and usually a horse.

Mr A.H. (Harry) Ashford was born in 1925 on Bulford Camp, where his father managed the privately owned Pitt's Garage from when it was built in 1920 until his death in 1947. He provides us with a fascinating insight into life on the garrison during the 1930s. Before managing the garage, his father had served for 18 years in the Army, four of them in France during the war. Along with his silver watch chain and French watch, he would wear his medal ribbons on his waistcoat, including the 1914 Star sometimes referred to as the Mons Star, to indicate to his military clientele that he had previously been one of them. The 1914 Star had been awarded to the 'Old Contemptibles', the soldiers of the original British Expeditionary Force and their immediate reinforcements, who had fought in France and Belgium at the very start of the war.

The garage was owned by a gentleman called H. Norman Pitt, who had been an Army officer during the war and served in France in the Motor Transport section of the Army Service Corps. It was here that he met Mr Ashford's father, and two other sergeants called Harrington and Davenport. Back in England he established three garages, the main one being in Amesbury, which was managed with him by Harrington, while Davenport took charge of the smaller of the three at Larkhill.

The Bulford garage was always open, never shutting its doors before 7p.m., and as the family (husband, wife, three sons and three daughters) lived in a house attached to its back, all the soldiers on the garrison knew where to go to get a can of petrol, or some paraffin for their lamps. Mr Ashford also ran a taxi service, which he drove himself in the early hours to pick up officers from the train station at Salisbury. When they reached the age of 17, his two eldest sons would also help to drive the taxi, beside working around the garage. In poor weather they might also pick up the midwife at night and ferry her to a civilian home; at all other times she rode her bicycle. Most people still had battery-powered wirelesses, so they also offered a battery charging service, which the whole camp made use of, both officers and men. There were very few electric wirelesses in the 1920s and early 1930s. The battery chargers ran 24 hours a day, and were checked at 10p.m. and again at 7a.m. Distilled water would be added if required and charged batteries would be removed in order that another could take its place.

The garage had three petrol pumps selling National Benzol, Shell and Shell Mex. Shell Mex was the cheapest at 1s. 1½d. per gallon. Oil was another big seller due to the poor quality of the engines at the time, and water was used by the gallon to fill overheated radiators. As a young lad, Harry would cash in by offering a water service to drivers from his half-gallon watering can, and charging them a farthing if they required a top up.

The Tidworth Tattoo was one of the garage's busiest periods as, for three evenings each year, people would drive from miles around to be there. More vehicles would pass through Bulford on their way up the Tidworth road than at any other time of year, and Mr Ashford would sit on Church Bank counting them as they passed, later recalling that hundreds would go by. The garage would stay open until after midnight, as cars returning south via Bulford later that night would stop to buy petrol and oil. Because the returning traffic was often very slow, vehicles also needed topping up with water.

The garage catered mainly for officers, who were the only soldiers that could afford to buy and run motor cars, although other ranks might appear with motor bikes. It also repaired bicycles, and did a big trade in tyre tubes, lamps and wheel spokes. It decoked engines, replaced big ends, fitted new piston rings, changed wheels, and serviced engine parts.

> 'There was always a car over the pit,' explains Mr Ashford, 'and all parts were washed in paraffin, so they looked almost new when they were collected and the bill was made out. All officers and other ranks paid in cash, unless they were personally known to Dad and were usually over the rank of Major. He didn't trust cheques, even from officers, and hence he was respected for his

> honest ways, never over-charging anybody. Quite often the family gave cups of tea to officers in their kitchen as they waited. Generals such as Wavell and Freyberg knew my mum well, as did the three local policemen and a few RSMs. They all knew that my Irish mother always had the kettle on the range.'

'Salisbury Plain was full of skylarks, hares and thousands of rabbits,' continues Mr Ashford. 'Before the war my brother Tom would drive the lorry to be loaded with hundreds of dead rabbits that had been gassed on the ranges.' Rabbit holes led to a few deaths and serious injuries, because horses put their feet into them. This affected not only officers and their mounts, but gun carriages being pulled by teams of six. There were usually five troopers with such a team, one sitting with each pair of horses, and the other two on the tender that held the shells. An accident caused by a rabbit hole could be devastating, and rabbits were considered a huge pest at the time. Mr Ashford can recall the funerals of soldiers that had been killed while exercising on the downs, their coffins drawn to the cemetery on gun carriages draped in the Union Jack, while hundreds of marching men would follow.

The garrison was serviced by several permanent shops, although there were very few civilians living in Bulford by then. Most of those that did remain lived in Sling Cottages and worked for the Army. The Barnes family ran the café, hairdresser and sweet shop, while the Petiffers looked after the dairy, and the Grist family the paper shop and tobacconist. There were also several lock-ups in Bond Street, occupied by Light's sweetshop, Bollons the greengrocer, a chemist, Hobbins the dress shop, a general store, and a small branch of WH Smith. Private tradesmen who visited the camp during the 1930s, with their horse-drawn carts and, later, motor vans, included Lakes the baker, Noyce for general van sales, Shrewton Dairies for milk, and Worsdall the greengrocers for fruit and vegetables. Of course, the Army had its NAAFI shops, but the civilian population wasn't generally permitted into these. The local butcher was Jones of Ludgershall, with whom Mr Ashford's mother, who kept chickens and ducks, would sometimes swap dressed chickens for good cuts of roasting meat. The local farm on Beacon Hill was run by the Godfrey family, which supplied Mr Ashford's mother with bacon joints; in return, they took paraffin to the farm which had no electricity supply.

Harry himself was a bit of a young entrepreneur, whose various paper rounds for Mr Grist had netted him £250 by the time he was 16 in 1941 when you could buy a house for £500. Most of the papers were sold to the soldiers on camp. With his father's persuasion he eventually joined the NAAFI, for which he was paid 12s. 6d. a week. When he wasn't cutting bacon, cheese or lard, he would be unloading lorries with sacks of dried peas weighing 240 lbs, or potatoes in one hundredweight sacks. The sides of bacon were usually around 50 lbs each.

He also served in the NAAFI bar two nights a week from 8p.m. to 11p.m., for which he was paid an extra £1 a night. When he was 17½ in 1943 he joined the Fleet Air Arm as a mechanic. After the war he returned to the NAAFI and either worked in, or managed, clubs as far afield as Vienna, Hamburg, Chippenham, Glasgow and Salisbury, before finally doing special duties in Cyprus in 1959, when the terrorist troubles were at their height.

The building of a complete new hutted camp at Bulford which was named Carter Barracks after Major General Sir Evan Carter, KCMG, CB, MVO, was also started in early 1939. This was one of the measures put in place to accommodate soldiers following the introduction of conscription in May 1939. The site chosen for these barracks, which were demolished again in 1978, had been a tented camp on the north side of the village towards Sheep Bridge.

For a while Sling Camp itself was divided into Gallipoli Lines and Baghdad Lines, but in April 1938 these were replaced by six separate barracks: Beacon Barracks, Gordon Barracks, Kiwi Barracks, Marlborough Barracks, Ward Barracks and Wing Barracks. In 1961, Beacon Barracks would be renamed Picton Barracks, and Marlborough Barracks would cease to exist, being amalgamated into the others around it.

Also around this time, the land that had been purchased to the south of Ludgershall towards Perham Down was first put to good use. Up until 1939, it had only been used for tented camps, at a place called Hedge-end Camp, but now a Mobilisation Depot and No. 1 Impressed Vehicle Camp was established near Hedge-end Plantation. At Tidworth, Mathew Barracks were started shortly afterwards, to accommodate conscripted soldiers after May 1939. These barracks, later demolished and now replaced by the Mathew Housing Estate, were named after Major General Sir Charles Mathew.

Further Acquisitions and Imber

Although during the war the Army had been omnipresent in the villages around Warminster, things soon returned to normal. Villages like Codford and Sutton Veny were able to resume their usual peacetime existences. The Army had also built camps all over the western side of Salisbury Plain, but with the end of hostilities they were obliged to vacate these sites as soon as possible. However, the War Office had come to recognise the potential of this area, and plans were put in place to increase its holding to the west of the Tilshead-West Lavington Road (A360).

The next phase of purchasing took place between 1927 and 1933, and as well as acquiring land and numerous farms, most controversially it meant the purchase of the entire village of Imber. The inhabitants of this ancient rural settlement were evicted from their homes during the Second World War, but were never permitted to return.

52 *Although Imber Court was destroyed by fire, during renovations after it had been purchased by Thomas Holloway, it was quickly restored to its former glory.*

During the first war, Imber Court and other local farms had been used to billet soldiers. Although the estate was the home of Tom Dean and his family, it was actually owned by the Dean and Chapter of Salisbury Cathedral. A major change took place in the community when, in 1920, the ecclesiastical commissioners sold Imber Court to the Holloway family, the owners of the nearby Lavington Estate who had gained their fortune running a successful building firm in London. During renovations after the purchase, Imber Court was dramatically destroyed by fire believed to have been started by a workman's torch. To the Holloways, the estate was a financial investment and they had no intention of living at the Court themselves, so following its reconstruction it was managed for them by Robert Whistler, a retired Army major.

Most properties in the village were owned by a group of farming families and leased to those who worked for them, but these were difficult times for farming communities, as during the 1920s and 1930s there was an agricultural depression. Very low prices were being paid for grain, wool or mutton, all of which sustained Imber's traditional economy. Because of the unpredictability of the area's weather and its remoteness from markets, large-scale vegetable cultivation was impossible, and cattle were unsuitable for life on the downs. All these factors contributed to the overall decline in local land prices.

The War Office approached the local landowners individually, with offers that were difficult to ignore. If a farmer sold his land to the War Office, it was guaranteed to be rented back to them at very favourable rates. Farmers were also told that, during peacetime, the land would only be required by the Army for a couple of weeks a year to accommodate an annual exercise which would only take place after the local harvest had been yielded. It was also feared that, if it eventually came to compulsory purchase, the price paid would be much lower. In his book *Little Imber on the Down*, Rex Sawyer writes: 'The tenancy agreements offered to the Imber farmers varied but were generally considered quite reasonable. The Hoopers at Brown's Farm, for example, were offered £20 an acre. Stock-proof fencing, gates and other maintenance work were also financed by taxpayers' money.'

As with earlier purchases to the east, the land around Imber would be considered either Schedule I or Schedule III. Schedule I land was rented at 10s. an acre per annum to the farmer and compensation given for any damage done by the Army, whereas Schedule III land was rented at 2s. 6d. per acre but with no compensation in the event of damage.

So, by 1932, with the exception of the church, the vicarage and the school, which were still owned by the church commissioners, the entire village of Imber had been sold to the War Office by the landowning families. Little did they realise that,

53 *This photograph of Imber was taken around 1935. In the foreground is a little bridge known as Blacksmith's Arch, upon which stands a young Ken Mitchell. The blacksmith's shop is just out of the picture on the right, although the war memorial, erected in 1920, can be seen. Also in the picture are St Giles' church and, behind Ken, the roof of Parsonage Farm. Listed on the memorial were the names of all those from the village who served in the Great War, and the three who died: Ernest Marsh and Harold Kitley, killed in France, and Arthur Norris, who died at Gallipoli.*

54 *Sir Alan Brooke was the Commandant of the School of Artillery from 1929 to 1932. He was also the garrison commander, and was largely responsible for the transformation of Larkhill at that time. He is pictured here during a Southern Command anti-invasion exercise in the Second World War.*

in the years to come, they would be indiscriminately evicted from their homes along with the rest of the village.

Although the War Office now owned almost a complete village, its real purpose had been to acquire the land that surrounded it, much of which had been used during the war for artillery practice. In the parishes of Imber and West Lavington alone, this amounted to 935 acres. But acquisitions continued, and between 1927 and 1939 a total of 27,158 acres was added to the War Office estate on Salisbury Plain, much of which would later provide land for Imber ranges and the School of Infantry at Warminster.

Understandably, in the villages around Imber, events were followed with keen interest in case a similar fate should befall those communities. At the village of Chitterne, a few miles to the south-east, people became more than a little concerned. In *Chitterne – a Wiltshire Village*, Sue Robinson writes:

> When the farmers in the Warminster area were notified that the War Department still needed large stretches of land, it was thought they would soon acquire the whole of Chitterne. *The Times* newspaper noted that there were 'serious complaints from archaeologists and agriculturists'. And at the Warminster Urban Council meeting the Water Committee reported that 50,000 gallons of water were wanted by the military for 3 months.
>
> However, these initial fears that the entire parish of Chitterne would soon fall prey to military take-over, as Imber had done, proved to be unfounded. Nevertheless, by 1937 the War Department had compulsorily purchased all of R.J. Farquharson's estates and land, amounting to 3,000 acres, as well as Manor Farm and Elm Farm buildings and lands, and all the outlying field barn settlements. The three farms in the old Chitterne St Mary parish, The Manor, Clump Farm and Glebe farm remained in private hands.
>
> In the village itself farm cottages now owned by the WD were numbered separately from other village dwellings and local men were employed by the Defence Property Services Agency to look after them. The agency had a depot at the former farmyard of Elm Farm.

Other War Office purchases at this time included Ladywell Farm, Edington, Boreham Farm, Bishopstow, and Manor Farm, Upton Scudamore while, on the eastern side of the Plain, 3,000 acres were purchased at Everleigh and Collingbourne Ducis at a cost of £46,850. The 1,100-acre Everleigh Manor Estate was finally acquired in 1954 for the cost of £21,000.

The School of Artillery

Although several areas of Salisbury Plain had been used as artillery ranges since the turn of the 20th century, it was not until after the First World War that Larkhill became almost entirely a gunner preserve. The School of Artillery opened in July 1920, with the closure and then amalgamation of several other schools, the School of Gunnery having been transferred from Shoeburyness to Larkhill in 1914. Then, in 1915, the Schools of Instruction for Royal Horse and Royal Field Artillery were also transferred to Larkhill. In 1920, these were amalgamated with the Overseas Artillery School opened on Chapperton Down, in 1916 to standardise the terminology of its students.

Although the land at Chapperton Down, which is now part of the Imber Ranges, had been used extensively in the war, it was given back to agriculture in 1919 until such time as the War Office purchased it. In the meantime, the Chapperton Down Artillery School moved to Larkhill where it was renamed the School of Artillery, and became the foundation block for the current establishment. Its staff was enlarged by recruiting instructors from other artillery schools, and since then Larkhill and gunners have become almost synonymous.

The new School of Artillery opened under the directorship of Brigadier General H.W. Newcombe, CMG, DSO, who had been the last commandant of the Chapperton Down Artillery School. Before long his command increased to include the Heavy Artillery Gunnery School and the Siege Artillery School which had previously been at Lydd.

At the time of the school's inauguration it had four core objectives: to qualify field artillery instructors so that they might assist with the training of both regular and territorial units; to develop and write all artillery

55 *The Ordnance QF 25-pounder gun-howitzer was successfully developed at Larkhill between the wars, and became legendary because of its rugged dependability. This gun is manned by members of Y Troop, D Battery, the Honourable Artillery Company. Far left is crime writer Michael Gilbert, who served with the HAC before, during and after the War. He took part in the very first post-war TA training camp at Larkhill in 1947.*

training manuals and other literature; to carry out trials of any new guns and shells that were being developed, as well as any other equipment associated with the artillery role, and, finally, become the stage for fire-power demonstrations.

The presence of the school brought about the transformation of Larkhill itself, which included the building of married quarters for both officers and men on the south of Fargo Road. This area became known as Strangways and was named after Brigadier General T. Fox Strangways, killed at Inkerman in 1854 while commanding the Royal Artillery during the Crimean War. It was now that The Packway, the rough chalk track across the downs trodden by shepherds and packmen for centuries, became the main route through a thriving garrison town.

Although the construction of Strangways had begun in 1928, the credit for Larkhill's transformation is largely attributed to the garrison commander, Brigadier A.F. Brooke, DSO, who was also the commandant of the School of Artillery from 1929 to 1932. With the help of his adjutant, Captain J.H.W.G. Richards, and the War Office Land Agent at the time, Lieutenant Colonel H.L. Piggot, OBE, the general appearance of this bleak outpost was greatly enhanced. Many of the disused wartime huts were knocked down, the roads improved and trees planted to give the location a more inviting feel. Brooke went on to command the 2nd Army Corps of the British Expeditionary Force in 1939, and was later promoted to Field Marshal, and Chief of the Imperial General Staff.

At Larkhill the lessons of the previous war were studied and techniques for the next devised. The main piece of field artillery at the time was the Ordnance QF 18-pounder. Variants of this weapon were in widespread service before the Second World War with British and Commonwealth troops, but many would be left behind on the beaches of Dunkirk and those that survived used largely for training and home defence. In terms of medium and heavy artillery, the QF 4.5 inch howitzer Mk 2 was another ageing relic captured in quantity in France and afterwards used for training among the British, although some did see further service in the deserts of Africa. The Germans turned some of these captured guns against Britain, employing them for coastal defence purposes.

56 *A gun crew loading a 25-pounder gun-howitzer during a Southern Command divisional artillery exercise at Larkhill.*

At Larkhill the Ordnance QF 25-pounder gun-howitzer was successfully developed

57 The Royal Artillery (Bulford) Harriers were re-established in 1919, and in 1934 new kennels were provided at the northern end of New Ward Road, where they were later converted from harriers to foxhounds, and their title changed to the Royal Artillery Hunt. This painting by Nikki Moore, from 2007, is titled 'Drawing the Westdown Gorses, RA Hunt' and was commissioned by the Royal Artillery Hunt at the time of the retirement of their master Nick Foster.

during the inter-war years. This versatile gun was first accepted into service after a trial at Larkhill, shortly before the outbreak of the war, and became legendary because of its rugged dependability. It was used in a variety of terrains and for various roles, and became the main piece of British field artillery for much of the war.

A showpiece of Larkhill during the inter-war years was the Hamilton Battery, which was made up of two 9.2 inch howitzers along with one 8 inch gun and one 6 inch gun. The Larkhill Military Railway passed very close to the battery, and from it ammunition would be transferred via a narrow-gauge tramway which led behind the guns. Although the howitzers were set up on semi-permanent emplacements, the guns were loaded on wheeled carriages and could be moved if required for ceremonial purposes. The battery was named after General Sir Ian Hamilton, and was situated close to the quarters of the Royal Artillery's 1st Survey Company, where crockery was known to bounce across the table every time the guns were fired.

During the 1930s, gunners on the Plain weren't restricted solely to the School of Artillery. Many still resided at Bulford, where in 1917 the pack of harriers known as The Royal Artillery (Bulford) Harriers had been put down. The pack was re-established in 1919, and in 1934 new kennels were provided at the northern end of New Ward Road, where they were later converted from harriers to foxhounds and their title changed to The Royal Artillery Hunt. The hunt is still going today, and is the last surviving pack of military foxhounds in the United Kingdom, although now totally self-financing. Perham Down, for a short period between 1921 and 1922, also became the home of the Anti-Aircraft Artillery School, before it moved to Biggin Hill, leaving Perham Down available for the next great development in warfare, the age of the tank.

Tanks at Perham Down

Just as Larkhill would become principally associated with artillery, Perham Down became known for tanks. On 26 April 1922 the 5th Battalion The Royal Tank Corps arrived at the camp, and remained there until 1939.

At the time of the Armistice there were 18 tank battalions in France and a further seven in England, but these were soon reduced to five, all of which were based either at Bovington or Wareham. Perham Down had become hutted in 1915, but the accommodation and facilities certainly didn't suit a tank battalion: the nickname of 'Perishing Down' indicates the esteem in which soldiers held it. So, in 1922, a budget of £32,500 was awarded to transform the site.

By 1927 the camp had been considerably altered, including the erection of several tank hangars. The existing barracks, although still hutted and made of wood and corrugated iron, were suitably converted and brick-built married quarters and houses for officers were constructed. When finished, the camp could house a battalion of 492 men and 26 officers. Having said that, the old hutted camp could once accommodate 4,878 men and 160 officers.

Quite probably, tanks were experimented with on Salisbury Plain during the First World War, no doubt using Larkhill to test their weapons. According to Rex Sawyer, there were rumours of 'mysterious mobile metal boxes' being encountered by local people on downland tracks. He also suggests that prototype military vehicles were actually made at the Bratton Iron Works using large water containers, or 'tanks'.

Just as the RAF worked alongside the Royal Artillery to direct their gunfire, by 1924 the same principle had been adopted by the Royal Tank Corps. This was demonstrated to Lieutenant General Sir Walter Congreve, VC, KCB, during a mock battle staged on Perham Down on 17 May of that year. Observing the demonstration from Warren Hill, he saw a battle in which two RAF aircraft bombed enemy gun positions, on to which they afterwards directed the fire of the tanks. As well as the 5th Battalion The Royal Tank Corps, other units taking part in this demonstration were 1st Brigade Pack Artillery, Royal Garrison Artillery, a battalion of the Wiltshire Regiment and a battalion of the King's Shropshire Light Infantry. Coincidentally, this was the year in which the Royal Tank Corps adopted their famous black beret. Up until then, like the majority of British servicemen, tank soldiers still wore the traditional peaked field service cap. However, inside a tank, especially when using a gunsight, the peaks were found to be totally impractical. The beret was officially adopted in March 1924, so at the demonstration at Perham Down in May they would presumably have been on view.

The idea of forming a tank brigade was experimented with on Salisbury Plain as early as 1921, and by May 1927 the Experimental Mechanised Force came into being, which included tanks working in conjunction with infantry, machine-gunners and artillery. Winston Churchill, Chancellor of the Exchequer at the time, attended a demonstration on Salisbury Plain in September 1927 of this force, which included medium tanks belonging to 5th Battalion Royal Tank Corps from Perham Down.

The 1st Tank Brigade was formed on 1 April 1934 under the command of Brigadier P.C.S. Hobart, DSO, OBE, MC. With a headquarters at Pinehurst Barracks, Farnborough, his brigade consisted of four tank battalions: 1st (Light) Tank Battalion at Perham Down, 2nd Battalion at Farnborough, 3rd Battalion at Lydd, and 5th Battalion also at Perham Down. However, experimentation with three of these battalions had begun as early as 1931 when, in September of that year, the entire brigade trained at Tilshead. Further summer camps would be held at Tilshead during the coming years, with brigades spending about six weeks at a time training there under canvas. They would also go to Lulworth camp on the Dorset coast each year around late March to early April for gunnery practice, also under canvas.

After the formation of the 1st Tank Brigade in 1934, Perham Down saw a period of considerable change. New brick barracks were built to house the men of the 5th Battalion, and the 1st (Light) Tank Battalion moved temporarily into the huts vacated while new barracks were built for all. In 1936 the new barracks were finished and a firework display was arranged by the building contractors to mark the event. The new barracks were named Cambrai Barracks and Cachy Barracks and, in similar vein, new roads in the garrison area were given names associated with tank engagements of the First World War: these included Cachy, Flers, Fremincourt, Kemmel, Drocourt, Fontaine and Bermicourt. Somme Road connected Perham Down with its near neighbour Ludgershall. As well as officers' quarters built to the west of Cachy Barracks, there were four houses for senior garrison officers named Bourlon House, Hamel House, Cachy House and Cambrai House. Each barrack also had messes for officers, warrant officers and sergeants.

There were infant and senior schools for the children situated in the married quarters of this growing tank town, but initially church attendees would have to march back and forth to Tidworth Garrison. The highlight of the school year was the Christmas Party organised by members of the Royal Tank Corps. A highlight of the year for the adults was the Cambrai Ball, attended by both officers and men. This was always held on the Friday or Saturday nearest to Cambrai Day, which celebrated the first great battle in which tanks took part on 20 November 1917 (although it was actually at Flers on 15 September 1916 that tanks were used for the very first time). There was also a polo ground situated where the current sports field stands and, while training took place at Tilshead, a families day would always be arranged, coaches bringing wives and children to the camp from Perham Down for a day of sports and fun. Each of the barracks also had welfare institutes for the men.

58 *This postcard, titled 'Salisbury Plain Camp 1938', shows a private in the Wiltshire Regiment.*

In May 1937 the 1st Tank Brigade Headquarters moved from Farnborough to Perham Down as part of the 3rd Division, and for a short period the garrison enjoyed relative stability. But war clouds were gathering again and things were set to change. Two completely new camps built at Perham Down were Fowler Barracks, named after Lieutenant General Sir John Fowler, and Busigny Barracks, named after a French village.

On 18 March 1938, the 1st (Light) Battalion left Cachy Barracks and entrained at Ludgershall on their way to Southampton and, from there, Egypt. Their place at Perham was taken by the newly reformed 8th Battalion Royal Tank Corps, who vacated the camp along with the 5th Battalion in 1940, leaving Perham Down to the 54th Training Regiment Royal Armoured Corps. In order to accommodate the training regiment, the 5th Battalion went under canvas at Windmill Hill, where they remained for two months as they were brought up to full wartime establishment with the arrival of reservists. They also received some new tanks, notably a few A9 and A13 Cruisers to complement their ageing Squadrons. In 1939, the Royal Armoured Corps was created, which effectively saw the end of the Royal Tank Corps, whose soldiers would become members of the Royal Tank Regiment.

Warminster becomes a Garrison Town

By 1930 much of the downland to the north and east of the little market town of Warminster had been purchased by the War Office. The land was considered ideal for both summer camps and tank training. By the mid-1930s units of the Royal Tank Corps were training during the summer months on land to the north of today's military garrison but, unlike Tidworth, which came to dominate and replace existing villages, the town and garrison at Warminster have remained separate, so that the old Saxon settlement still retains much of its original identity.

From the very outset, Warminster garrison was developed with the Royal Tank Corps in mind, in order to provide an area in which greater tactical use of tanks could be made. In 1934 the planning for Boreham Camp was in hand, on the north side of Woodcock Road, and in February 1937 work began on the building of two complete barracks: Swinton Barracks and Elles Barracks would be able to accommodate a complete tank battalion each. Also under construction at this time were the buildings required for the headquarters of a tank brigade, most likely intended for the 1st Tank Brigade which had been relocated to Perham Down in May 1937.

Swinton Barracks were named after Major General Sir Ernest Swinton, KBE, CB, 'Father of the Tank Corps'. Swinton was an officer in the Royal Engineers who went to France in 1914 as a war correspondent, although he was also Assistant Secretary of the Committee of Imperial Defence at the time. After seeing the difficulties of operating any form of transport in the trench-torn Western Front, he put forward a proposal to develop a tracked armoured vehicle that could traverse broken ground using caterpillar tracks from American Holt farm tractors. Winston Churchill, First Lord of the Admiralty at that time, was so impressed with the idea that he appropriated the funds to enable Swinton's ideas to develop.

Elles Barracks took their name from Major General Sir Hugh Elles, KCMG, CB, DSO, who was the first tank commander in the field, the first tanks being crewed by men of the Heavy Section of the Machine-Gun Corps. This section went to France in August 1916 with their first complement of tanks, and on 29 September Colonel Hugh Elles, another officer in the Royal Engineers, took command.

The accommodation at these barracks included the first examples of the 'Sandhurst Block'. These newly designed units consisted of a central section with four wings, all of which were centrally heated throughout. On the ground floor of the central portion were kitchens and dining rooms; elsewhere were sleeping quarters and sitting rooms, and each floor had its own lavatories, bathrooms, drying rooms and rooms in which equipment could be cleaned. Compared with the wooden and corrugated iron barracks of previous decades, these must have been comfortable and very popular with the soldiers.

The first of these barracks to be completed were Swinton Barracks, which were officially opened on 23 September 1938 by the Rt Hon. Leslie Hore-Belisha MP, the Secretary of State for War. By the time of the opening ceremony, the

59 *Vickers Medium Tanks speed across Salisbury Plain in the late 1930s. This was the type of tank used by the 5th Battalion Royal Tank Corps at Perham Down and the 2nd Battalion Royal Tank Corps, who were under canvas near Warminster. By the outbreak of war they were becoming out-dated.*

60 *Members of the Royal Tank Corps manoeuvring model tanks on a model battlefield, somewhere in England.*

3rd Battalion Royal Tank Corps had already moved into their new homes, having previously been stationed at Lydd in Kent. Many people from the town descended on the barracks to find the entrance decorated with colourful bunting, streamers and pennants, while the flag of the 3rd Battalion fluttered above the administrative block. The Secretary of State was greeted, and the battalion welcomed to the town by the Chairman of Warminster Urban District Council, Mr E.G. Durbin.

As the work on Elles Barracks continued, the 2nd Battalion Royal Tank Corps, who were its intended occupants, found themselves under canvas on the downs; they were still there when war was declared in September 1939. In the meantime, 89 married quarters for non-commissioned ranks were built at Firbank Crescent, and 23 quarters for married officers at Oxendene.

The presence of tanks also meant that changes were needed to the roads along which they regularly drove. In *Chitterne – a Wiltshire Village*, Sue Robinson writes:

> Where previously the village roads had been 'bad and exposed', now the War Department strengthened some of them with concrete under the tarmac, for heavier Army traffic. Road signs appeared alongside village roads warning of tanks, or later banning tanks from certain roads. Tank crossings, where tank tracks crossed roads, were marked by two rows of old railway sleepers topped by bright yellow caps. Some of these have recently been replaced by metal posts, but still with yellow caps.

Warminster had hardly become accustomed to its new residents when the outbreak of war changed everything. Suddenly, the 2nd and 3rd Battalions had to prepare for active service, and the camp at Warminster would be taken over by another training regiment for the Royal Armoured Corps. Whereas Perham Down had become 54th Training Regiment, Warminster became 57th Training Regiment Royal Armoured Corps. There would eventually be 11 such establishments around the country, the others being at Catterick, Bovington and Farnborough, and the 53rd, 59th, 60th and 61st Training Regiments all based at Tidworth. So the headquarters of 1st Tank Brigade never actually moved to Warminster before the war, and all battalions were soon to be in action,

many equipped with either Matilda Mark I or Matilda Mark II Infantry tanks.

At the training regiments at Warminster, Perham Down and Tidworth, recruits would undergo a 24-week course, the first eight of which were taken up with general military or basic training, including drill, skill at arms, driving, PT, gas training, elementary tactics and navigation. For the rest of the period they were trained to take their place in the Royal Armoured Corps, which needed drivers, gunners, wireless operators, mechanics, fitters, electricians, clerks and storemen. In many ways, the Royal Armoured Corps was one of the most independent and diverse arms of the Army, and Salisbury Plain was its foremost training ground.

RAF Activities and Army Co-operation

After the First World War, all flying instructors were trained at the Central Flying School, Upavon, which was now an RAF base. The expansion of the school required substantial numbers of civilian workers, and in around 1920 council houses were built for them at Avon Square, on the Andover Road, between the village and the airfield itself. Upavon, like many of Salisbury Plain's villages, was growing in size. A Church of England chapel was opened on the station in 1921, later dedicated to St Peter, and by 1936 a Roman Catholic chapel, dedicated to St Thomas More, was in use.

Several further changes occurred at the station during the inter-war years. In 1924, as well as being the home of the Central Flying School, Upavon became an operational airfield when No. 3 Fighter Squadron moved in. Two years later, in 1926, when No. 17 Fighter Squadron arrived to join it, the decision was taken to move the Central Flying School to Wittering in Northamptonshire. In 1934 the fighter squadrons left, and night-time flying techniques were developed here, the new training base for a number of squadrons belonging to the Fleet Air Arm: 800, 801, 820 and 821. The Central Flying School returned to Upavon on 2 September 1935, and in 1942 was replaced by No. 7 Flying Instructors' School, many Royal Air Force servicemen passing through Upavon in the years to come. HM King George VI honoured the Central Flying School with a visit on 5 May 1938.

Netheravon became a Flying Training School in July 1919, and from December of the same year was titled No. 1 Flying Training School, continuing as such until 1931 when it moved to Leuchars. Then, until January 1933, No. 509 Flight of the Fleet Air Arm was formed at Netheravon, to train Royal Navy pilots. But Netheravon also continued to be the site of numerous trials and experiments. For instance, it was here that the first military rotary-wing flying trials took place in 1933; it was the venue for the world record free-fall parachute jump from 20,000 feet; and in 1937 a new height record was set by an aircraft climbing to over 53,000 feet over Salisbury Plain. In 1939, Netheravon would be transformed into a Naval Advanced Flying Training School for Fleet Air Arm pilots, which continued to operate there until April 1942.

Boscombe Down was also re-opened during the inter-war years as a Royal Air Force bomber base. The necessary land was purchased in 1927, and the building began of what would become an important permanent aerodrome, opening in 1930. It's also worth noting that in 1938 a new airfield was started in Hampshire at Middle Wallop, which isn't actually on Salisbury Plain but it deserves a mention because Royal Air Force and later Army Air Corps pilots from here, especially in the helicopter age, would use the Plain as their main training ground.

In the early days of military flight, the Royal Flying Corp's loyalty was to the Army, and one of its main aims was to direct artillery fire. After the war there were calls for the Army to have its own aircraft again, as the Royal Navy had been permitted to do. With the formation of the Royal Air Force, however, it was inevitable that the main interest of the new service would be in training and developing fighter and bomber squadrons and Army Co-operation became less and less important. Things like reconnaissance, photography, message dropping and direction of artillery fire were given low priority.

In 1920 a School of Army Co-operation was established at RAF Old Sarum, and from 1925, during annual manoeuvres, pilots belonging to these squadrons, flying an assortment of Bristol Bulldogs, Wallaces, Audax and Hector aircraft,

61 *Cierva C30 Autogyros were used in Army Co-operation squadrons in the mid-1930s. Trials in directing artillery fire from C30s were conducted at Larkhill in 1935.*

62 *Westland Lysander Army Co-operation monoplanes flying over Hampshire in the late 1930s.*

used a landing ground about a mile to the west of Tilshead. Also in 1920, the Royal Air Force School of Balloon Training was established at Rollestone. Battery commanders and gunner officers from the School of Artillery at Larkhill were able to see how balloons and kites, as well as some aircraft, could help direct their fire. But Royal Air Force higher command didn't know quite what to do about Army Co-operation as, over the next few years, the school at Rollestone changed its name and ethos on several occasions. In 1929 it was the Larkhill Kite Balloon Station, in 1932 the RAF Balloon Centre and, in 1936, No. 2 Balloon Training Unit.

During the 1930s, several imaginative Royal Artillery officers with an interest in aviation came to understand that the only way to control artillery fire from the air effectively was for the Army to have its own pilots. The problem would be convincing higher authorities. Several of these officers had qualified as RAF Army Co-operation pilots, and together they formed the Royal Artillery Flying Club at Larkhill, where they carried out surreptitious trials using light club aircraft and voice radio to spot for and control the guns on the ground.

Most influential amongst these pilots would be Captain (later Major General) Hetman Jack Parham, CB, CBE, DSO, and Captain (later Lieutenant Colonel) Henry Charles Bazeley, DSO. Parham learned to fly on autogyros in 1934 and then bought his own light aircraft, an Aeronca, for which he built his own airstrip and hangar near Bordon in Hampshire. Bazeley was seconded to the RAF in 1936 for Army Co-operation duties. From then on he campaigned energetically for Air Observation Post (AOP) light unarmed spotter aircraft to be flown by Royal Artillery officers based in a field next to the guns themselves and so capable of providing a quick reaction. After many disappointments he would later command No. 651 Squadron on active service in North Africa. This was the very first AOP unit, and was formed at Old Sarum.

In 1933 an article appeared in the journal of the Royal United Services Institute, written by Captain Parham, in which he criticised the existing method of directing artillery fire from Army Co-operation aircraft: aircraft sent morse code signals over a one-way radio, and responses were by means of ground signals. Even though many of the pilots were Army officers seconded to the RAF, they were serving in RAF squadrons and as such were detached from the guns. They would be briefed for each sortie at an airfield some distance from the front line, which meant there was always a considerable delay until the target could be engaged. As far as Parham was concerned, the answer to this problem was obvious: if the aircraft and pilots were attached to the Royal Artillery they could operate from a field right next to the gun batteries, and by using two-way radios, their response to enemy fire could be immediate.

Parham might have been the most outspoken member of the Royal Artillery Flying Club but, as its secretary, Charles Bazeley was the man to get things done. The club spent time and energy holding trials at Larkhill to see if their plans could work, but Bazeley knew the scheme would only ever be adopted by the Royal Air Force after official trials. With persistence, he persuaded the War Office with Air Ministry agreement to hold official trials in 1939 using Westland Lysanders, the aircraft used by the existing RAF Army Co-operation squadrons, and the trials proved a great success, the only drawback being the fact that in a combat scenario the unarmed Lysander could be very vulnerable to the attention of enemy fighter aircraft. These fears proved to have substance in May 1940, when the Army Co-operation units encountered Messerschmitt Bf 109Es in France.

However, through the commitment of Parham and Bazeley, the Second World War would see great advances in army aviation as it broke away from the Royal Air Force, first with the Air Observation Post squadrons and, finally, with the formation of the Army Air Corps.

Chapter Four

The Second World War

The outbreak of war in September 1939 brought an end to the building of any more permanent barracks in the garrison towns of Salisbury Plain. However, the vast increase in the size of the Army, due to both volunteers and conscription, led to an outcrop of hutted camps throughout the area similar to 1914. Once again, soldiers of every description began to pour on to the Plain in their thousands, preparing to fight in the deserts of North Africa, the jungles of Burma or the fields of France.

Over the next six years, men from so many different countries trained or assembled here for the forthcoming campaigns that the garrisons on Salisbury Plain were likened to 'the United Nations'. When the Americans arrived in 1942, to prepare for Operation Overlord, General Eisenhower described it as 'the best training ground in the United Kingdom'. From garrison towns and temporary camps on the Wiltshire downs, this united army of nations prepared itself and set off to liberate the Nazi-occupied lands of north-west Europe.

A United Army of Nations

For the duration of the war, Larkhill was occupied almost entirely by the Royal Artillery, or units that provided the regiment with technical support. There were of course exceptions to this rule but, in the main, the soldiers who passed through the garrison were either British or foreign gunners attending the School of Artillery. Here they took part in courses that mainly employed either the 25-pounder gun-howitzer for field artillery, or the 4.5 inch howitzer for medium and heavy artillery. Other weapons used for training included the heavier 9.2 inch howitzer and the 12 inch howitzer. The school expanded quite dramatically at this time as its responsibilities significantly increased, with many more courses and staff members being required. In 1939, there were fewer than 80 instructors, but this had risen to more than 250 by 1943, with a total of 2,000 staff. Ultimately, the establishment could cater for 590 officers and 700 other ranks at any given time.

63 *King George VI and Queen Elizabeth visit Larkhill on 3 April 1941. The roar of the guns causes the Queen to cover her ears.*

To accommodate all the people passing through Larkhill, which included numerous conscripts and officer cadets, new hutted camps were desperately needed. The first of these was Newcombe Lines, named after Major General H.W. Newcombe, CMG, DSO, who had been the first commandant of the School of Artillery. Newcombe Lines was situated to the south of Fargo Road, between Willoughby Road and Alanbrooke Road. By 1943, the Fargo Hutted Camp had been established in the area between The Packway and Fargo Road, to the west of Willoughby Road, but in November 1944 Fargo Hutted Camp was one of several establishments chosen for conversion into prisoner-of-war camps, ready to accept German servicemen captured on the continental battlefields. The School of Artillery and the Larkhill ranges received their first official royal visit when His Majesty King George VI, accompanied by Queen Elizabeth, paid them a visit on 3 April 1941.

Before the war, both Royal Artillery and Royal Engineer officers were trained at the Royal Military Academy at Woolwich in London, while cavalry and infantry officers attended the Royal Military Academy at Sandhurst in Surrey. After the war started, the work of both these academies was suspended in favour of sending potential officers to Officer Cadet Training Units, or OCTUs. These were established at many locations to train officer recruits for their various regimental roles. The Royal Artillery established five OCTUs, where potential officers learned the skills they would require to become battery or troop commanders. The courses usually lasted between four and six months, although most cadets would have already undergone some training in training regiments and, from 1942, in primary selection centres. Some also underwent several weeks of pre-OCTU training.

From September 1941 onwards, No. 122 OCTU at Larkhill was particularly aimed at candidates for the anti-tank role. The other artillery OCTUs were at Aldershot, Catterick, Llandrindod Wells and Filey, which was later moved to Ilkley, the Aldershot OCTU later moving to Alton. There was also a Royal Artillery Advanced Survey Regiment at Shrivenham in Wiltshire, and in 1943 Larkhill would also become the location of the 4th Royal Artillery Training Regiment (Field).

At the OCTU there were usually about fifty cadets in a troop occupying two barrack rooms. During the course the students took part in marching drills, gun drills, driving, lectures and panoramas. The first exam, taken after two weeks, was on military law, with other tests following in quick succession. The spring course of 1940 proved particularly eventful as it coincided with the evacuation of Dunkirk, and the students were required to vacate their barrack rooms in order that they might be used to accommodate soldiers returning from France, who arrived in trucks from the railway station. Then, over the next few moonlit nights, they were given rifles and live ammunition and set up at positions near Stonehenge to prepare for a suspected landing by German parachute troops.

During one of the First Aid lectures at the spring 1940 course, an instructing officer was reported to have said, 'He'll be no good in action', after one of his cadets fainted while a description was being given of how to apply a tourniquet. The cadet in question was a certain George Ward Gunn, who within 18 months had won the Victoria Cross at Sidi Rezegh in North Africa as a second lieutenant serving with the 3rd Regiment, Royal Horse Artillery. On 21 November 1941, Gunn commanded a troop of four anti-tank guns which came under heavy attack from German tanks. After three of his guns were knocked out and the fourth damaged, and with most of the crews dead, he continued to sight the gun while a sergeant loaded it. They fired off between 40 and 50 rounds, crippling at least two enemy tanks, before Gunn was himself killed.

64 *'Trenches in the chalk – behind the 122 OCTU RA at Larkhill.' This picture was painted by the artist Bruce Allsopp in 1941, while training at Larkhill's Officer Cadet Training Unit.*

As well as being a principal training area for British troops, Larkhill would once again be used by the men of several allied armies to perfect their skills. Although the Canadians were housed in permanent barracks at Aldershot this time around, gunners of the 1st Canadian Division, who arrived in Britain in April 1940, would attend the School of Artillery to receive instruction on the 25-pounder gun-howitzer.

Gunners from Newfoundland would also train at Larkhill. The 57th (Newfoundland) Heavy Regiment spent a week at Tilshead in February 1941, where they fired 25 pounders and 9.2 inch howitzers on the West Down ranges. They were later converted from heavy to field artillery and re-named 166th (Newfoundland) Field Regiment RA and, as such, were stationed at Larkhill between 17 November 1941 and 4 May 1942.

The 59th (Newfoundland) Heavy Regiment remained in the heavy role and trained at Larkhill in July 1941, where they used 12 inch howitzers. In March 1942 they were also on West Down, this time practising with 6 inch guns and 9.2 inch howitzers. Between April and June 1943 they took part in various formation exercises at both Larkhill and West Down. Australian and New Zealand personnel would also attend gunnery courses at the School of Artillery, although neither of these nations had troops stationed around the garrison. The New Zealanders were based around Aldershot and the Australians around Bulford and Lopcombe Corner.

With the outbreak of war, the 2nd and 3rd Battalions of the Royal Tank Regiment had moved out of their new homes at Warminster to make way for the 57th Training Regiment Royal Armoured Corps. A Royal Army Ordnance Corps workshop was established to the north of the Great Western Railway line which was known as the Royal Army Ordnance Corps Depot Warminster. Its main purpose was to maintain the tanks and equipment of the 2nd and 3rd Battalions, and when they deployed to France with the British Expeditionary Force a field workshop from the depot accompanied them. The remaining staff continued to operate the workshop until after Dunkirk.

It was realised after the German army's spectacular success in France and Belgium that more units needed to be equipped with armour, and more soldiers trained to fight in armoured vehicles. So, as well as becoming

65 *One of the Newfoundland artillery regiments training with the 9.2 inch howitzer at Larkhill.*

66 *During the Second World War the Australian Imperial Forces on Salisbury Plain were mainly based at Lopcombe Corner Camp. In this photograph Australian soldiers are seen returning to camp after a route march through the Wiltshire countryside.*

the home of 54th Training Regiment Royal Armoured Corps, Perham Down was selected as the site of No. 103 OCTU, through which potential officers for cavalry and armoured units would pass, one of several OCTUs dealing with the armoured role. The newly formed Royal Electrical and Mechanical Engineers took over part of the Royal Army Ordnance Depot at Tidworth when this corps was formed in 1942.

The 1st Guards Armoured Division was formed in September 1941 as part of this mechanised programme, and the Royal Army Ordnance Corps workshop at Warminster became the 1st Guards Armoured Division workshop. Various battalions within the division trained on Salisbury Plain with Crusaders or Cruisers. For instance, the 2nd (Armoured) Battalion, Welsh Guards, was initially equipped with Cruiser tanks which had previously belonged to the Royal Tank Regiment, but after being stationed at Codford between 5 September and 31 December 1941 they were upgraded to Covenanter IIIs in March 1942. By the time the Guards Armoured Division came to see action, in Normandy in 1944, they had been equipped with either Sherman, Cromwell or Churchill tanks, and at the close of hostilities they relinquished their tanks and reverted to the infantry role.

In 1943, before Warminster was handed over to the Americans, it was the base for the 42nd Armoured Division. The armoured elements of the division were stationed at Warminster itself, while the infantry were in a hutted camp at Longbridge Deverill. At

67 *Amesbury Abbey, which in 1940 became the headquarters of the Australian 18th Brigade.*

this time, the division took part in a semi-documentary film commissioned by the War Office, entitled *The Way Ahead*, which starred many well-known actors of the day, including David Niven. It was directed by Carol Reed and written by the combined talents of Peter Ustinov and Eric Ambler, the latter of whom was serving in the Royal Artillery and became Assistant Director of Army Cinematography at the War Office.

David Niven, by this time a huge star in Hollywood, had already served in the British Army before the war, and had re-enlisted at the onset. At the time the film was made, he was serving as an officer in the Rifle Brigade at the rank of Lieutenant Colonel. Set in North Africa at the time of El Alamein, the film was conceived by Niven himself, who was of course happy to participate in its production as it was intended to be viewed as a promotional exercise for the armed forces. Among those taking part, and appearing for roughly seven seconds of screen time, was the 1st Battalion Highland Light Infantry, for which officers were paid £3 a day and other ranks £1, but whereas the other ranks were permitted to keep their money, the officers had to give theirs to regimental funds.

Longbridge Deverill Camp was one of several temporary sites in the Warminster area, initially used by the British Army and later occupied by the Americans. There were two camps at Sutton Veny, Long Iver Camp, which was situated on the Deverill road, and a smaller one in Best's Lane. There was also a tented camp at Bishopstrow, and once again Codford was used to house numerous troops, with the main camps being on the road from Codford St Peter to Manor Farm, and between Codford St Mary church and Little Wood. Codford became home to several elements of the Guards Armoured Division from September 1941 to 1943, whose troops proved very popular with the local residents.

The Australian Imperial Forces on Salisbury Plain were mainly based at Lopcombe Corner Camp, about eight miles south of Tidworth, from their arrival in June 1940 until their departure in October of the same year. Initially, the camp was occupied by the 18th Brigade, with its Force Headquarters at Amesbury Abbey, but they were later joined by the 25th Brigade. Arriving as they did in the critical summer of 1940, 18th Brigade was given the task of being Southern Command's Striking Force against any possible invasion attempt. Southern Command Headquarters had moved to Wilton House, just to the north of Salisbury. The Australians also established an infantry leaders' school and, of course, gunners attended courses at Larkhill.

At Bulford, some Australians went under canvas next to Carter Barracks, while the Royal Australian Army Ordnance Corps occupied buildings on the west side of the Amesbury Road, and a detachment of the Australian Army Provost Corps was stationed in Church Street. Lopcombe Corner Camp was visited by His Majesty King George VI on 4 July 1940, and the Prime Minister Winston Churchill on 4 September 1940, when they watched Australian troops exercising in the field, and marching on the parade ground. The Australians eventually left Britain for the Middle East between November 1940 and January 1941.

Also at Bulford was a Mule Company of the Royal Indian Army Service Corps, camped on the north side of the sports ground between New Ward Road and Sling Road in June 1940. They also occupied adjoining woodland, and probably arrived here from Amesbury station after Dunkirk. Many other units arrived on the Plain at this time, including a group of around 200 French soldiers remembered by Harry Ashford. He recalls that whereas most of the British troops arriving by train or lorry looked dishevelled and unkempt, with no rifles, kit bags or even hats, the Frenchmen had all their equipment intact. They were approached to join the Free French Army, but refused and asked to be returned to France. The Royal Navy sent them by destroyer further down the French coast, where they landed before the arrival of the Germans. As far as they were concerned, the war was over.

Many women also served in the various branches of the armed forces, so naturally they, too, would have experience of Salisbury Plain. In particular, ladies of the Auxiliary Territorial Service found themselves attached to garrisons and camps all over the area. Barbara Mottershead was stationed at Larkhill, from September 1939 until April 1943, and Winterbourne Gunner, until December 1944. From this eventful period of her life she recalls the following:

> While at Larkhill garrison I was with the 122 OCTU and when it moved to become 122 A/T OCTU (in Kent) I was transferred to the School of Artillery. When we first arrived at Larkhill the small number of ATS girls were housed in the redundant 'Tin' garrison church. Later we were put in various tin huts dotted around the camp, before we were moved into the main camp. When I was moved to the School of Artillery we were housed in spider blocks. At Winterbourne Gunner I was stationed at the Army School of Chemical Warfare and the ATS were housed in what had previously been the married quarters.

68 *Viola Cove, who joined the WAAF in 1943, after working in the Pewsey Council Office. Her sister Evelyn Joan Dancey explains that in early 1945 Viola was based at RAF Upavon, where she worked as an instrument repairer on the aircraft. At that time the American Army would send transport to take the girls to their dances at Tidworth. Viola also served at Innsworth, Pembrey, Melksham and Henlow, which is where she met her husband-to-be.*

To cater for the welfare of all the soldiers passing through Bulford, the Biscuit Trade Association kindly provided the funding for a new YMCA building to be established next to the cinema, opposite the junction of Haig Road and Marlborough Road. This became known as 'The Biscuit Hut', and next to it another hut was provided by the Young Women's Christian Association, recognising the fact that many women now served in the garrison, especially those belonging to the Auxiliary Territorial Service. At Tidworth, the garrison theatre was taken over by the Entertainment's National Services Association and, as well as staging concerts, this was the location of several rousing lectures by great military figures. These included Field Marshal Earl Alexander of Tunis, KG, GCB, GCVO, DSO, MC, and Field Marshal Viscount Montgomery of Alamein, KG, GCB, DSO, DL. No doubt, all ranks and nations represented in the audience were suitably inspired.

In the Wiltshire Sky

RAF Upavon and RAF Netheravon continued in their training roles, although in March 1942 the Central Flying School at Upavon changed its name to No. 7 Flying Instructions School. At Old Sarum the School of Army Co-operation continued to train pilots and observers in artillery spotting, reconnaissance, air photography and bombing. At RAF Boscombe Down the bomber squadrons moved out and the Aeroplane & Armament Experimental Establishment moved in and, in similar vein, in 1941 a High Altitude Flight was formed. These would also be joined by the Test Pilots' Training School, and an Intensive Flying Development Flight, established in order to assess the long-term reliability of aircraft.

As well as being the site of experimentation and development, numerous operational squadrons were also based at Boscombe Down during the war, including No. 249 Squadron, who found themselves here at the height of the Battle of Britain and were charged with patrolling the area west of Southampton. On 16 August 1940, while flying from Boscombe Down, Flight Lieutenant James Nicolson won – incredibly – Fighter Command's only Victoria Cross of the war. He received the award because, even though his own aircraft had been set on fire by a Messerschmitt Bf 110 and he had suffered terrible injuries, Nicolson remained in the cockpit, only baling out once he had pursued and shot down

69 *Austers in their many variants were used by the Air OP squadrons. This is an Auster Air OP Mark 7 used for training at Old Sarum.*

the enemy aircraft. Dozens of young fighter pilots went 'above and beyond' the call of duty on a daily basis during the summer of 1940 but he is the only one thus rewarded.

Between the wars, thanks to the efforts of Captains Parham and Bazeley, as well as several other individuals at Larkhill, the idea of using Army pilots to spot for artillery had been taken more and more seriously, especially after the success of the official trials in 1939. When the war began, the Air Ministry reluctantly agreed to provide light aircraft to equip an experimental Air Operations Flight, called D Flight, that went to France in April 1940. However, before it was able to prove itself in battle, the German attack terminated all such activities. Back in England the Air Ministry sought to wind up the enterprise, but Captain Bazeley persisted, taking D Flight to every available formation exercise in order, to demonstrate the techniques he and his pilots had evolved. The aircraft of D Flight could be positioned in a field close to the guns and their skilled pilots fly below treetop height, from where they were able to pop up to observe the fall of shot and enemy positions before diving back into cover again.

Eventually the Air Ministry agreed to the formation of several Air Observation Post squadrons, the first of which, commanded by Bazeley himself, was No. 651 Squadron, formed at Old Sarum on 1 August 1941. This was also the first unit to see action, after deploying to Tunisia in November 1942. In all, 16 squadrons were formed, including three Canadian and one Polish, many of which were equipped with various versions of the Taylor-craft Plus aircraft, followed by the Auster I and III. In 1944 and 1945 the Air Observation Post squadrons were re-equipped with the Auster IV and V, both of which had an American Lycoming engine.

Although the pilots for these squadrons were drawn from the Army, and almost exclusively from the Royal Artillery, the aircraft were owned and maintained by the Royal Air Force, which often led to a conflict of interests. Potential pilots were initially attached to the Royal Air Force for elementary flying at RAF Peterborough, where they were taught to fly Tiger Moths. From there

70 *A Hotspur glider flying over Wiltshire. On 16 April 1942, Winston Churchill was given a demonstration at Netheravon of airborne troops being towed to their landing zone in Hotspur gliders pulled by nine Hector aircraft.*

they were transferred to the Air OP Course at 1242 Flight RAF Larkhill, before finally joining their respective unit. The 16 squadrons were numbered from 651-66, with all of the following being formed at Old Sarum at various times: 651, 652, 653, 654, 655, 658, 660, 661 and 662. Another three were formed at nearby RAF Andover. In December 1944, due to Royal Navy involvement, the School of Army Co-operation at Old Sarum became the School of Air Support, and it remained so until May 1947, when it was renamed the School of Land/Air Warfare.

In 1940, several small airfields were once again established on Salisbury Plain, one of which was at Shrewton, between Shrewton Farm, Elston Hill, The Bustard and Greenlands Farm Camp. From July 1940 it was used as a relief landing ground for Netheravon, and later by Army Co-operation units and gliders. A little to the south, another area near Rollestone between the B3086 and A360 was used as the Royal Air Force Anti-Gas School, where gas attacks made from aircraft were simulated. The Balloon Training Unit had occupied Rollestone Camp previously, and when the Anti-Gas School moved out, in 1945, the airfield was used as a landing ground for the Auster V aircraft of No. 657 Air Observation Post Squadron. The Royal Air Force finally relinquished control of the land at Rollestone to the Army in 1946.

Another wartime airfield was situated at High Post, approximately five miles north of Salisbury, on the west side of the A345 opposite the High Post Golf Club. There was already a private airfield here before the war, which had opened in May 1931 and was used by the Wiltshire Light Aeroplane & Country Club. This later became the Wiltshire School of Flying. In 1935, a series of new aerodrome buildings were constructed, which included offices, an observation tower, a wireless room and hotel with a fully licensed restaurant. All of this made High Post very popular with inter-war aviation enthusiasts so, when the war began, the potential of the site was immediately recognised by the Air Ministry, who took it over and initially designated it a practice landing ground for aircraft flying out of RAF Old Sarum. But, from 6 June 1940, the airfield was used by No. 112 (AC) Squadron of the Royal Canadian Air Force. The hotel, which is still trading today as *The Inn at High Post*, became the officers' mess.

When the Canadians departed, both RAF Netheravon and RAF Upavon made use of High Post as a site for flying training. Then, Vickers Supermarine started using one of the hangars to assemble Spitfires. Their main factory in Southampton had been bombed, so the various components were built at garages in Salisbury and Trowbridge before being brought to High Post for assembly. Later, and with the agreement of the American Army Air Force, this Spitfire production line and test flying facility was transferred to the airfield at Keevil, just to the east of Trowbridge, where the Ministry of Aircraft Production built a hangar. After the war, High Post airfield returned to private use, but finally closed in February 1947. The hangar that was used for assembling Spitfires is now occupied by a company called Apsley Engineering.

There were several other small airfields, including a site to the north of Everleigh used as a maintenance dispersal area and an offshoot for the Central Flying School. New Zealand Farm, on Littleton Down, on the edge of the current Imber ranges, was established in October 1940 for training pilots from Upavon in night flying. It would later become an outstation for Boscombe Down. Another airstrip a mile to the west of Tilshead was opened in June 1940 and used by the Lysander II aircraft of No. 225 Squadron, who had moved there from Old Sarum on 1 July 1940. The men were accommodated in tents at Tilshead Lodge Camp, which adjoined the airfield, but when the weather was particularly bad they were known to use Tilshead Lodge itself. Between June 1941 and May 1946, a small airfield at Oatlands Hill, near Winterbourne Stoke, was also used to train Army Co-operation Command pilots. And aircraft flying from airfields like Thruxton, Andover, Middle Wallop, and Chilbolton in Hampshire would also use the skies over Salisbury Plain for both training and operational purposes.

71 *Officer in an Air Observation Post squadron, c.1944, with an Auster V shown in the background. From a painting by Alix Baker PCAFAS FRSA.*

Airborne Forces

From 1940, a number of imaginative innovations were tested in aerial warfare, following the example set by the successes of the German Army in the early stages of the war. Among these was the use of parachute and glider-borne troops, who could arrive directly on to a battlefield. In June 1940, Winston Churchill himself gave

73 *Paratroopers boarding converted Whitley bombers at Netheravon, ready for a practice drop.*

the order to raise a complement of 5,000 British airborne troops, and once again Salisbury Plain would play a crucial role in their development.

In June 1940, the Central Landing School had been established at RAF Ringway near Manchester, where much of the technique for parachuting and carriage of troops by glider was developed. The parachute troops would be Army personnel, while the pilots who ferried them into battle would belong to specially designated Royal Air Force squadrons. Having said that, it was decided that the pilots of the actual gliders, towed behind the powered aircraft, could come from the Army because once they were on the ground their job as aviators was effectively complete. So, on 21 November 1940, 11 Special Air Service Battalion was formed, in which 66 Army pilots were attached to Army Co-operation Command for preliminary training before converting to gliders. The battalion had a glider wing and a parachute wing. On 21 December 1941 the glider wing became 1st Glider Regiment, which in turn led to the formation of the Army Air Corps on 24 February 1942. The parachute wing became the 1st Battalion The Parachute Regiment on 1 August 1942.

The 1st Airborne Division was formed under the command of Major General Frederick Browning, GCVO, KBE, CB, DSO, with his headquarters at Syrencote House at Milston, about two miles north of Bulford, where he took residence between November 1941 and May 1943. On 15 January 1942, No. 38 Wing of RAF Army Co-operation Command was formed under the command of Group Captain Sir Nigel Norman, with a headquarters at Netheravon. The job of No. 38 Group would be to lift the troops of 1st Airborne Division into the air and deliver them to their targets. The group's aircraft would need to play dual roles: they would carry fully equipped parachutists to a point above their dropping zone, and they required the ability to tow gliders full of troops which, when released, would be steered by Army pilots to their landing sites. Netheravon ceased to provide the Royal Air Force with a base for flight training, concentrating instead on the Army Co-operation role. Around the same time, the Naval Advanced Flying Training School at Netheravon, which had been there since the beginning of the war, was also closed. But this had more to do with the fact that arrangements had been made to train Fleet Air Arm pilots in the United States rather than in Britain.

No. 38 Wing comprised four squadrons, Nos 295-8. No. 295 Squadron was formed at Netheravon on 3 August 1942, and remained there until moving to Holmsley South in May 1943. No. 296 Squadron was formed at Ringway as the Glider Exercise Unit, and moved to Netheravon in February 1942 but left again for Hurn in Dorset in July 1942. No. 297 Squadron was formed at Netheravon on 22 January 1942, as the Parachute Exercise Squadron, and also moved to Hurn in June 1942. And, finally, No. 298 Squadron operated out of Thruxton before moving to Tarrant Rushton in November 1943. While exercising over Salisbury Plain, all these squadrons could be seen flying Whitley V aircraft, although the entire group would eventually progress to more suitable aircraft such as the Halifax and Albemarle.

In March 1942, the 1st Parachute Brigade, a component of the 1st Airborne Division, arrived at Kiwi Barracks, Bulford, under the command of Lieutenant Colonel Ernest Down. Geoff Baker was a member of the 1st Parachute Brigade and has a particular reason for recalling his time at Bulford:

> In 1942 I arrived at Bulford Barracks to join the Parachute Regiment. Later my fiancée came and lodged with Mrs Plank in a nearby village and we met whenever it was possible. I jumped once on to Salisbury Plain from a Whitley bomber which had an aperture in the floor and we had to swing our legs into this to jump. If the next man was too keen, he put his boots into the face of the previous man. This happened to me and I had to turn up for our date with a swollen face. In September, on my 21st birthday, we did a practice drop on the Plain and afterwards I met up with her at Stonehenge. She brought with her a picnic and my presents. We sat on one of the surrounding stones and used it as a table. It was a lovely autumn day and nobody could have wished for a better 21st.

On 16 April 1942, Winston Churchill was given a demonstration at Netheravon during which parachute troops jumped out of 12 Whitley bombers belonging to No. 38 Wing, while other airborne troops were towed to the landing zone in Hotspur gliders pulled by nine Hector aircraft. Detachments from Nos 296 and 297 Squadrons were inspected by Their Majesties The King and Queen at Netheravon on 21 May 1942, while on 1 April of the following year the King made an unaccompanied return visit to the airfield.

By now Tilshead airfield had become the Training Depot of the 1st Battalion, the Glider Pilot Regiment, under the command of Lieutenant Colonel J.F. Rock, formerly of the Royal Engineers, his second-in-command being Major G.J.S. Chatterton. These men were responsible for transforming their recruits, into not just glider pilots but infantrymen as well, as once the gliders were on the ground the pilots would have to become members of the airborne fighting force and fully trained infantry soldiers. As most of the volunteer pilots had already served in infantry battalions, and many were veterans of earlier campaigns, the transition was relatively painless, although the discipline at Tilshead was known for being unnecessarily strict. While No. 296 Squadron, the Glider Exercise Unit, was based at Netheravon, it also maintained an outstation at Shrewton airfield and, once fully trained, the glider pilots would join the 1st Glider Regiment at Airborne Camp, Tilshead, from where they would begin to exercise with the airborne troops.

In June 1942, 21st Independent Parachute Company arrived at Airborne Camp but was transferred to Newcombe Lines, Larkhill, in the following October. The Glider Pilot Regiment was itself eventually moved out of Airborne Camp to Fargo Camp, in September 1943, being replaced at Tilshead by an armoured regiment. Shortly afterwards the 6th Airborne Division was formed for operations in north-west Europe, and once again Syrencote House was used as the division's headquarters, commanded from May 1943 by Major General R.N. Gale, OBE, MC, and from here the airborne participation on D-Day was planned. In October 1943, No. 38 Wing RAF Army Co-operation Command was renamed No. 38 Group, and by February 1944 this consisted of ten squadrons, flying a mixture of Albemarles, Halifaxes and Stirlings.

Three weeks before D-Day, on 19 May 1944, both Major General Browning and Major General Gale were prominent figures when the King and Queen visited Netheravon, accompanied by Her Royal Highness The Princess Elizabeth. Also in attendance that day was the Air Officer Commanding No. 38 Group, Air Vice Marshal L.N. Hollinghurst, CB, OBE, DFC, and the Air Officer Commanding-in-Chief of the Allied Expeditionary Air Force, Air Marshal Sir Trafford Leigh-Mallory, KCB, DSO.

73 *Paratroopers boarding converted Whitley bombers at Netheravon, ready for a practice drop.*

British airborne troops would famously take part in many important wartime operations, including Operation Overlord in June 1944, Market Garden in September 1944, and Operation Varsity in March 1945, by which time many of the airborne troops were ferried in Dakota aircraft or Horsa gliders towed by Stirling bombers. Although none of these operations was actually flown from Salisbury Plain, it was here that the men who took part originally trained for these historic occasions. At the entrance to Syrencote House is a plaque featuring a Pegasus emblem, a sign now evocative of British airborne troops. The inscription reads:

> This house was used between November 1941 and May 1943 as the Headquarters of Major General, later Lieutenant General F.M. (Boy) Browning, CB, DSO, the founder of Airborne Forces. Much of the planning which led to the formation of the 1st and 6th Airborne Divisions was done in this house. In May 1943 it became the Headquarters of the 6th Airborne Division commanded by Major General R.N. Gale, OBE, MC, later General Sir Richard Gale, GCB, KBE, DSO, MC. The airborne assault by the 6th Airborne Division on the night of 5/6 June 1944 to secure the left flank of the Allied invasion of Europe – Operation Overlord – was planned and mounted from this house. Ad Unum Omnes.

The War comes to Wiltshire

Because the Wiltshire downs were used so extensively by the military forces of different nations, the war affected the people of Salisbury Plain in many different ways. The area would often be a target for German attacks and Larkhill experienced a number of these, the worst occurring on 17 January 1941 when several bombs fell on Newcombe Lines in the early hours of the morning, killing 19 people and wounding many more. Several of the dead belonged to a recently arrived civilian intake who were sleeping in huts on the west side of the camp, about eight of which were destroyed. Damage was also caused to the gun park, workshops and garages.

The only enemy attack on the airfield at Netheravon occurred on 27 July 1942, when machine-gun bullets tore into the officers' quarters and those occupied by members of the Women's Auxiliary Air Force, causing superficial damage to the buildings but, thankfully, no injuries to personnel. Across the Hampshire border at Middle Wallop, the worst attack took place on 14 August 1941, when six people were killed, 15 injured and a hangar was demolished.

74 *Staff Sergeant, Glider Pilot Regiment, 1944, with Horsa glider. From a painting by Alix Baker PCAFAS FRSA.*

75 *The Duchess of Marlborough on the right, supervising the unloading of rations at an ATS camp at Tidworth, where there were a thousand women under canvas.*

German aircraft would often be seen hunting across the Plain, looking for suitable targets, and Imber was targeted more than once. On one occasion incendiary bombs were scattered all around the village but, fortunately, no serious fires were started. One of the bombs was discovered some considerable time later in a thatched barn, lying among sheaves of corn as workers were taking them out for threshing. It had gone straight through the roof of the barn but failed to detonate. Had it done so the consequences could have been devastating, as the fire would not only have burned the corn, barn and other nearby buildings, but attracted the attention of further bombers to the village.

The small village of Chitterne was nearly bombed one Saturday night, when a dance was taking place in the village hut. Perhaps it was the lights generated by the occasion that prompted the German bomber to drop 21 bombs around the scene. 'The first one,' explains Sue Robinson, 'landed in the first big field on the left going towards Codford, the rest were dropped in an arc with the last one just missing the road to Warminster. One young girl at the dance was so scared that she fainted right away and had to be revived.' Sue Robinson also describes the lucky escape that some local boys had when a German aircraft mistook, presumably, a new farm building for a military establishment:

> Some village lads had a narrow escape from enemy guns when they were gathering piles of straw with a horse and trailer in farmer Robert Long's fields. A Heinkel 111 German plane suddenly flew in low with its machine-guns firing directly at them. Luckily no one was hit. At the time the Poldens had just had a shiny new roof put on their field shed and this was blamed for attracting the pilot. One of the bullets went clean through a pigsty and a pig's trough, the head of the bullet lodging in the opposite side of the trough, letting all the water out. The pilot was clearly seen by other village lads from their window in The Round House, and they felt sure he saw them too as he swooped to avoid the large chestnut tree in their garden. They were sure he would have fired at them save for his need to clear the tree. The plane was later shot down near New Zealand Farm at Gore Cross.

Like anywhere else in the country, even the most remote parts of Salisbury Plain had its own home front. Several men from the area departed on military service, but others took part in essential war work. Wiltshire men who paid the ultimate price in both world wars are in churchyards throughout the region, their graves often surmounted by distinctive War Graves Commission headstones. Food production was increased in the Imber valley and prisoners of war were sent to help with the harvest. Many evacuees arrived from London to escape the bombing, residing temporarily with local families. Also at Imber was a company of Local Defence Volunteers, later known as the Home Guard, commanded by Captain Arthur Williams, who lived at Brown's Farm. Williams had served on the nearby gunnery ranges during the First World War and married a local girl, so he knew the area well.

76 *The Prime Minister Winston Churchill watches a mock battle on Salisbury Plain, staged by Southern Command during the summer of 1941.*

A small group of soldiers was billeted in the village of Chitterne, some of them at Chitterne Lodge, which had been purchased by the War Office as part of R.J. Farquharson's estate. Others were put into rented accommodation, while both soldiers and evacuees from London were accommodated in the local schools. Chitterne also had a Home Guard platoon, which was commanded by Mr Limbrick of Manor Farm. Its headquarters was situated in an old shepherd's hut at the top of Shrewton Hill, alongside the track to Yarnbury. Chitterne also had Army Cadets, as Sue Robinson explains:

> It was decided that it would be a good idea for the newly formed guard to practise apprehending the enemy, in case of a future invasion. So the Chitterne cadets were given the task of pretending to be 'Germans'. They were told to go and hide in the village. The adults in the guard were to seek them out and apprehend them. But, so successful were the cadets at performing their roles in the exercise, that not one single 'German' was caught.
>
> Later, the very disgruntled and embarrassed Home Guard accused the cadets of cheating by hiding in trees. The crafty cadets had won the day by using their local knowledge. Some hid inside old hollow trees and others climbed up into the lime trees that bordered the Walk well out of sight of their elders.

Air Raid Precaution wardens also patrolled this remote swathe of downland looking for lights to put out. One of the wardens, Norris Hooper, who lived at Cornbury Farm near Gore Cross, was part of a detachment of 12 men based at West Lavington and was equipped with one tin helmet and one pair of size 9 Wellington boots. Norris would normally wear the helmet because his duties included stopping traffic at Gore Cross and it was regarded as a sign of his authority. Chitterne had at least two air-raid wardens, William Churchill and Thomas Gorry, whose job, at the first sign of an air raid, was to cycle from one end of the village to the other, blowing a trumpet. When the first incident occurred, this was done perfectly, although Sue Robinson explains that whoever was on duty at the time forgot to sound the all-clear afterwards, 'so, technically, Chitterne is still under curfew'.

One of the most tragic events of the entire war to take place on British soil happened on 13 April 1942, during a demonstration by the Royal Air Force, in conjunction with No. 5 Corps of Southern Command, designed to show the devastating effect that fighter aircraft equipped with cannons and machine guns could have on enemy convoys. It took place on the downs near Imber and, during it, 25 military personnel were killed. One of the inexperienced Hurricane pilots of No. 175 Squadron mistook a gathered crowd of military observers for a dummy column of soldiers, and fired directly into it. The demonstration was a rehearsal for a planned visit by Winston Churchill and General Marshall, Chief of Staff of the United States Army, which went ahead as planned three days later. The most senior officer at the rehearsal was Lieutenant General Sir Gerald Templer, GCB, GCMG, KBE, DSO.

Several factors contributed to the disaster. No. 175 Squadron had only formed at Warmwell in Dorset on 3 March 1942, just six weeks earlier. There was confusion over the radio telephone messages, and over the markings put down to distinguish the target from the observers. The pilots were informed that spectators would be completely surrounded by a white strip, but on the day the order was unclear and only enough material was available to form a dotted line at the front of the crowd. During the exercise the pilots flew into the sun, and the haze made it difficult for anything to be discerned. A 21-year-old Canadian, Sergeant William McLachlan, who had only been with the squadron for three weeks, fired a short burst directly into the crowd and only realised his mistake as people began to scatter in panic.

Some of the casualties were taken to Warminster Hospital, while others were ferried in a fleet of ambulances to Shaftesbury Military Hospital, where the medical staff worked for several days and nights. Of the 25 dead and 71 injured, many were high-ranking officers. Among the injured was Lieutenant the Hon. Robert Cecil, heir to Viscount Cranborne, who was Secretary of State for the Colonies.

As the war continued and training intensified, the position of Imber village became more and more untenable. It was right at the heart of an increasingly important training area and, although an exclusion zone had been placed around it of 1,000 yards, the safety of the inhabitants and their homes could

77 *Their Majesties The King and Queen observe a tank battle on Salisbury Plain, staged by Southern Command during the summer of 1941. The royal couple witnessed a demonstration of tanks and dive-bombers.*

78 *The Chitterne contingent of the Local Defence Volunteers, later renamed the Home Guard. This photograph was taken outside the front of Manor Farm at Chitterne All Saints. The commanding officer, Mr Limbrick, who lived at the farm, is seated in the centre. In the front row are the Chitterne Army Cadets.*

not be guaranteed. The situation was made even more acute when the American Army arrived on Salisbury Plain in 1943, with the 3rd Armored Division based around Warminster and the 4th Armored Division around Devizes. Imber ranges would feature extensively in their training schedules, so it was considered far too dangerous to allow residents to remain, not just in the village itself but also on the farms along what was known as the Imber Corridor. As the War Office had purchased the entire village, with the exception of the few properties belonging to the church, between the wars, it decided to exercise its power by evicting all the inhabitants.

On 1 November 1943 the Army called a meeting at the schoolroom to announce that the Americans would require the village in order to train for the type of street fighting they might encounter in the towns and villages of Normandy. The 135 residents were given until 17 December to leave their homes, a mere 47 days. The announcement took everyone by surprise; they had lived through military occupation during the First World War, so why should this be any different? Three days later a further meeting was held in the Town Hall at Devizes, where a fuller explanation was offered as to why the evacuation was necessary. The Army promised to help people find new homes if they were unable to do so for themselves, but everybody managed to sort something out. They also made a controversial promise that after the war the people of Imber would be able to return to their homes, although the War Office would later insist that no such promise was ever made. No legally binding document has ever been produced in support of the claim, but the residents unanimously agreed that a promise of some sort was given.

The remote rural community suffered the loss of their homes and way of life with great sadness. The blacksmith, Albert Nash, was so heartbroken that he died within a few weeks and was returned to St Giles churchyard for burial. Farmers were forced to sell their livestock at vastly reduced prices. By the time the war in Europe had ended the majority of Americans had already left the country, but the people of Imber, scattered across the county with relatives or in entirely new homes, were destined to remain exiles. They were informed that the War Office intended to retain the whole area as a permanent firing range and so re-occupation of the village by its former inhabitants was out of the question.

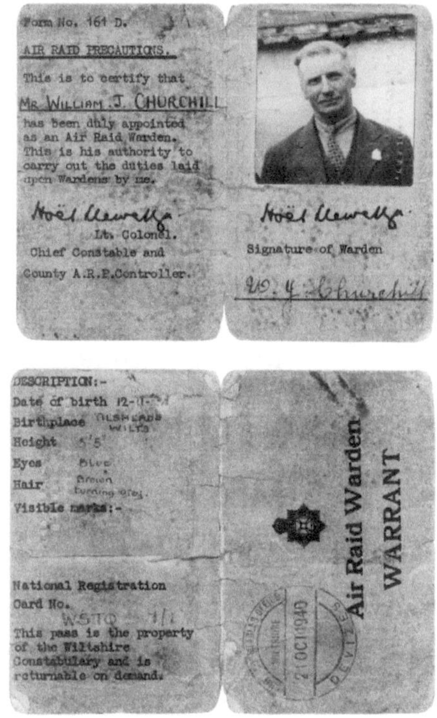

79 Warrant card issued to William Churchill, who was appointed an air-raid warden for the village of Chitterne in 1940.

Since the evictions, the only way an Imber resident can return to the village, like Albert Nash, is for burial in the churchyard. St Giles remained the property of the church, although important documents like the parish registers and churchwardens' accounts book were taken to Potterne church for safekeeping. In 1952 the Rous tombs were housed in Edington church, while elaborate woodwork went to a military chapel at Bratton and the pulpit was placed temporarily in the Army chapel at Warminster. In 1955 a group of villagers was given permission to return for a day, when a service held in the church was conducted by the Reverend Ralph Dudley, vicar of the newly combined parish of Edington and Imber. By 1960 these services, held on the Saturday nearest to St Giles's Day, had become an annual pilgrimage. Today, Imber is open to the public on various occasions during the year but the surviving residents are few and far between.

During the war the villages on the periphery of Salisbury Plain would again be visited by many different soldiers in the area to exercise on the downs. These soldiers would inevitably affect the lives of the people living there. The record of one village, Milton Lilbourne, is supplied by Mrs Evelyn Joan Dancey (née Cove), who lived at King Hall Farm from Easter 1935 until 4 December 1943 and kept a diary of the events. Milton Lilbourne is just to the east of Pewsey, close to the Pewsey White Horse, while Upavon, a few miles to the south-west, is where her late sister Viola Cove worked in early 1945 as an instrument repairer with the Women's Auxiliary Air Force.

Evelyn was working for Mr G.B. Nicol in Pewsey, who owned a shop selling all manner of things ranging from new and second-hand furniture to ladies', gents' and children's clothing, as well as household linen and the like. He was also a pigeon officer for Southern Command, and an army contractor who would supply Southern Command with certain services. For instance, in 1939 he was hiring out things like marquees, tents, rugs, tablecloths, serviettes, cookers, china, glass and cutlery to the various troops who came to Salisbury Plain for their two weeks' training. These were, of course, mostly territorials. Mr Nicol claimed he had enough glass and crockery to sit 300 people down to a seven-course dinner, all of which was left in two large huts behind the Post Office and another by the cinema. Using her diary as a reference, Evelyn provides the following:

> On 22 November 1939 the 7th Battalion Worcestershire Regiment came to stay at Milton. The officers were at Havering House, the sergeants at a cottage and the men in the stables at King Hall. This was during a very cold spell when every twig and branch were encased in ice. On the night of 16 December the soldiers were taken to Ogbourne to learn to dig trenches. The next day three of them came to our house for a warm. Colour Sergeant Carl Share asked if his wife could stay with us for Christmas. The men used to drill in the avenue leading to our house. They left the village en route to Marlborough then France on 14 January 1940 and I recall them singing 'We'll meet again' and 'Wish me luck as you wave me goodbye' while they were loading up by the church. After Dunkirk they went to the Far East and I later learnt Carl Share had been in the prison in Singapore.
>
> On 18 January 1940 B Company 4th Battalion East Lancashire Regiment came to Milton. We had six sergeants and their batman staying in our house. He didn't have much to do, so spent a lot of time playing his mouth organ 'South of the border'. They left on 20 February for Marlborough and A Company 4th Battalion East Lancashire Regiment arrived on the same day, and left again on 14 April, also for Marlborough.
>
> The Gloucesters were at Ogbourne and the late Duchess of Gloucester's brother came to the office. I once saw Amy Johnson in Pewsey, she was staying at the *Phoenix*. On 27 May 1940 Mr Nicol had an urgent request for all the mattresses he had, and as he wouldn't let them out without covers someone had to go to Shrewton laundry for some. Several of us girls stayed at work until 9 p.m. making covers out of a bale of calico, one cutting out and two sewing them on hand sewing machines. These were for soldiers returning from Dunkirk. On 21 September 1940, four of us cycled to Huish in our lunch hour to see David Niven and Primula Rollo married. I also recall seeing Freddie Mills the boxer in the tearoom in Pewsey High Street, where four of us girls went for our tea break. Several of us went with the Countess of Hardwick to help decorate the Bouverie Hall for a dance she was arranging for some of her husband's troops. She was living at Rockley Manor. In 1941/1942 there were New Zealanders in Savernake Forest. I used to meet one riding his bicycle into Pewsey as I was going home in the evening. He married Joan Lovell from the *Phoenix*.

80 *General Eisenhower visiting troops of the American 3rd Armored Division in February 1944, at Warminster.*

Mrs Dancey herself was called up in December 1943, but as the forces were full she was sent to work in a factory near Gloucester which made aeroplane propellers. She met her husband there in May 1945 after he had just returned from four years in the Middle East, after which she never returned to Milton Lilbourne. Her husband had been evacuated from Dunkirk on the paddleship the *Waverley*, which was sunk by the Germans.

The Americans

When the Americans entered the war in December 1941, virtually the whole of Salisbury Plain was given over for their use. From June 1942 the build up of General Omar Bradley's American First Army began, as tens of thousands of men trained in readiness for the liberation of Europe. Salisbury Plain was the ideal location for this training to take place, and Tidworth was established as the Headquarters of the United States II Corps, under the command of General Mark Wayne Clark, on 22 July 1942.

Consequently, of all the garrison towns, Tidworth probably saw the most American soldiers pass through it. So many, in fact, that as well as the barracks, they also took over the school and married quarters along Bazaar Road for extra billeting. The nearby camps at Park House, Tidworth Pennings, Windmill Hill and Lopcombe Corner were also given over for their use.

Over the next few years, various divisions of the First Army came and went. Initially, the 1st Infantry Division ('the Big Red One') under the command of Major General Terry Allen took up residence on 8 August 1942, remaining for

81 *General Eisenhower observes the use of a .45 cal. sub-caliber tank gunnery device at Warminster, during his inspection of the 3rd Armored Division in February 1944.*

the next three months with their headquarters in Lowa Road. Allen's assistant commander was Brigadier Theodore (Teddy) Roosevelt Junior.

Next to arrive, on 18 October 1942, were the 29th Infantry Division commanded by General Charles Gerhardt, who remained for seven months. This division was nicknamed the 'Blue and Gray', because it was made up of men from the states of Maryland and Virginia, which had opposed each other during the American Civil War. On 6 June 1944, the 1st and 29th Infantry Divisions, along with the 4th Infantry Division, would provide the main assault troops for the landings on Utah and Omaha beaches. After training on Salisbury Plain, they would attend the Assault Training Center at Woolacombe on the coast of north Devon.

Other units who passed through Tidworth prior to D-Day were the 2nd Armored Division, who arrived on 27 November 1943, and the 5th Infantry Division on 26 January 1944. After the invasion had taken place, these were followed by the 7th Armored Division, 9th Armored Division, 12th Armored Division and 8th Armored Division, most of which only remained for around two weeks before leaving for France on active service. Last to leave, on 3 January

1945, was the 8th Armored Division, and by 19 April 1946 the final supporting elements of the US force had vacated the garrison.

Wherever American soldiers went, ladies of the Red Cross would follow, to cater for their comfort and welfare. Red Cross clubs were established in every American camp, many of which were known as 'Donut Dugouts', because they would serve up fresh doughnuts to the troops every day. They also served hamburgers, coffee and soft drinks, but no alcohol. The men could relax, listen to music and read newspapers from home, and the staff would help them write letters to their loved ones, or just lend an ear if they wished to talk to somebody outside military circles.

In September 1942, Tedworth House became an American Red Cross club, opening its doors to the men in October. The club was managed by the wife of Brigadier Roosevelt and became very popular, with its doughnut-making machine and hamburger grill. Weekly dances were also held at Tedworth House and were a powerful draw for British servicewomen and local girls from Tidworth and the surrounding area. It is claimed that the 'Hokey-cokey' was introduced to Britain there in 1943. The Cadre Club was another spot in Tidworth where the Red Cross would organise dances which the GIs attended in the hope of meeting English girls. While the Americans were in town, Bob Hope and James Cagney were among the entertainers who came to the Tidworth garrison theatre.

At Perham Down, the 54th Training Regiment of the Royal Armoured Corps was moved to Catterick in order that elements of the 5th United States Armored Division could occupy the camp in February 1944. The 8th Cavalry Squadron and three artillery battalions remained there until leaving for France in June. At Ludgershall, the area to the south of the town that had been used by No. 1 Impressed Vehicle Camp, had also been used by a clothing sub depot, command ordnance depot, Royal Army Ordnance Corps workshop and Royal Australian Army Ordnance Corps workshop. In 1943, some of this land became a United States Army and Navy workshop, and here the American Army built three massive sheds in which vehicles were assembled and prepared to take part in the invasion of Europe. Jeep components would arrive in crates and, after assembly and checking, these would be driven out to their units. The Americans also built a railway link from this site to the Ludgershall-Tidworth branch line, in order to transport heavy equipment and vehicles,

82 *An American soldier training on Salisbury Plain. This was one of many GIs that Brenda Gower met at dances around the Andover/Tidworth area.*

83 *Staff at the American Red Cross Service Club on The Green in Devizes. Most of the girls who worked in these clubs were recruited locally.*

including complete tanks. After D-Day, some of this depot complex at the western end was taken over as a prisoner-of-war camp.

With the departure of 57th Training Regiment Royal Armoured Corps, also to Catterick, the barracks in Warminster were taken over by the United States 3rd Armored Division, commanded by Brigadier General Doyle O. Hickey, in September 1943. They also occupied the Guards Armoured Division workshop to the north of the Great Western Railway line, in order to maintain their own tanks and equipment. Elements of the division were scattered all around the Warminster area, with Combat Command B, the 33rd Armored Regiment, 391st and 67th Armored Field Artillery Battalions in the town itself; the 36th Armored Infantry Regiment at Long Iver Camp, Sutton Veny; the 83rd Armored Reconnaissance Battalion at Longbridge Deverill Camp; the 23rd Armored Engineer Battalion at Fonthill Bishop; the 703rd Tank Destroyer Battalion at Mere, and the 32nd Armored Regiment at Codford St Peter, with maintenance and supplies battalions in Codford St Mary. Stockton House was requisitioned for use by Combat Command A, 45th Armored Medical Battalion, and Trains Headquarters. Other divisional troops could be found at various locations in the county of Somerset, with the 54th Armored Field Artillery at Frome, the 143rd Armored Signal Company at Cucklington, Divisional Headquarters at Redlynch House, Bruton, and Division Rear at Wincanton.

Much of the division's training would take place at either West Down, or Bowl's Barrow to the south-west of Imber, and it was after they occupied Warminster that the decision was taken to develop the Imber ranges which led to the forced evictions at the village on 17 December 1943, and its subsequent use for training American soldiers in street fighting.

In December 1943, the 486th Armored Anti-Aircraft Battalion arrived at East Knoyle, to join the 3rd Armored Division. Manoeuvres on the downs were frequent and one in particular, supervised by General Hickey and his Combat Command 'A' staff from Stockton House, in which tank battalions exercised in conjunction with infantry, engineers and artillery, was held on land to the west of Chitterne. In early 1944, as the invasion of Europe drew closer, the manoeuvres became a weekly occurrence. At various times in those final months the division was visited by an impressive string of top brass, among them Montgomery, Eisenhower, General Bradley, Air Chief Marshal Tedder and the Duke of Gloucester. As the day of embarkation became more imminent, most of the units gathered together and set up tents on the Plain around Tidworth, waiting for the order to 'go'. D-Day occurred while the division still waited, then, finally, they struck their Wiltshire camps and headed to the ports of Southampton and Weymouth, arriving on Omaha Beach on 23 June.

Don R. Marsh served with the 143rd Armored Signal Company but belonged to the 3rd Armored Division, and had been sent with several others from his unit to join 142nd Armored Signal Company, 2nd Armored Division at Tidworth, just before they took part in the invasion. His original unit had been billeted in what was previously a mushroom farm at Cucklington, so the barracks at Tidworth were a welcome, but very short-lived, relief. His words are taken from an article published on the 3rd Armored Division's website:

> The First Soldier Len Mainiero called off our names one by one from his ever-present clipboard he habitually carried, handed our 201 Personnel folders over to a Corporal in Charge, turned and walked away without saying a word. It was over. My two well-traveled duffel bags were tossed in the back of the GI truck. I was ready for the short ride to our next destination – the Tidworth British Barracks, home of the 'Hell On Wheels' Second Armored Division. A small circle of old friends from the 143rd Signal gathered around for one last goodbye. After handshakes all around, I hopped in the back of the truck and snapped off a smart hand salute. Adios, amigos. See you in the Bulge.
>
> The trucks entered Highway 303 and wound through villages and small towns of Wincanton, Deptford and Amesbury, arriving at the British peacetime regular Army cavalry barracks at Tidworth on the Salisbury Plains. At the Main Gate, MPs checked the driver's trip ticket and passed us on through. We pulled up in front of the 142nd Signal Company Orderly where the Corporal with our records went in as we jumped down from the trucks.

84 *Mike Monez of the 4th Armored Division (Finance Group) pictured in 1943. He struck up a friendship with Devizes girl Elsie Lewis (née Caswell) who worked in the local Donut Dugout.*

> A call to the Wire Section and a Staff Sergeant Harry T. Hall came by soon to escort us to the wiremen's barracks. The two-storey red brick buildings – with windows – were a welcome sight after the six months at the Spartan-life 143rd Signal Company mushroom ranch. The Wire Section's second floor housed double-deck bunks with real, albeit 3-inch, GI mattresses! I hadn't seen those since we left Camp Shanks in October. Sergeant Hall told us to grab a bunk and settle in until he sent for us. A Corporal eventually came by to escort us to the Supply Room to draw two blankets for the sacks. No more roughing it on hard wooden boards – these were real GI cots with springs! Unfortunately, as luck would have it, we didn't get to make use of them very long. The trip to the battalion-size Mess Hall was a far different experience from the past six months in that the cooks had use of a bona fide kitchen instead of a makeshift field cook stove setup. The food was GI just the same. However, we ate out of our mess kits, as this still was the US Army. A trip to the shower room later was a shear pleasure – endless hot water, which we hadn't experienced since leaving the States. They had a movie theater with nightly showings. A small well-stocked PX was available. We were back in the real world once again with 'all the comforts of home' – but for how long?

Marsh had joined the 142nd Armored Signal Company commanded by Captain Henry J. Stuart, and was posted to the Combat Command A wire team. But he wouldn't be there for long. He had arrived on 24 May 1944, the invasion was imminent, and the entire camp had effectively been put on standby, with nobody going in or out. He continues:

85 *Alan J. Stewart from Belmont, Massachusetts, is pictured here at Tidworth with 14th Major Port headquarters.*

86 *An American soldier is seen here with Italian prisoners-of-war, in the village of Appleshaw, between Tidworth and Andover. The prisoners worked on local farms and usually roamed freely, but as many of the Americans were Italian-speaking, such as this GI, they often became friends.*

When we arrived at Tidworth, the whole division was already into a lock-down, and no one received any passes for leave or to go into the nearby towns. It was no secret the Second Armored Division would be one of several divisions included in the initial invasion forces. We were loaded and waiting for the word from General Eisenhower.

Now it was time to sit and wait. Sergeant Harry Hall assembled the wiremen and told us time was short – hours away, we would be on standby from here on in. We were then told to get a GI buzz-cut white-wall haircut and write 'the last letter home', which would be held for posting until we went out the gate heading for the docks of Southampton.

Parts of the general orders were 'do not keep envelopes containing our military postal address, unit references, no photos, and above all – no diaries'. The latter could have presented a slight problem for myself, but many of us totally ignored it. I stuffed the essential contraband in a compartment of the jeep. We were under wraps, so all shoulder patches had to be removed and the Division identification markings on vehicles had to be painted over.

While waiting, we were subjected to one more 'shakedown' inspection of our field packs. Anything unauthorised was tossed. Hiding forbidden items wasn't too difficult – we simply put them in our vehicles! The Company armorer, Cpl Vince Smedley, had issued me my 30 cal. carbine rifle, with extra ammunition magazines, which required a trip to the rifle range to zero-in the weapon. The live ammo distribution would wait until just before our convoy went out the gate en route to the dock area. The word came to 'Mount Up!' Our jeep and a small weapons-carrier wire truck joined the long parade of armored vehicles headed to the sea like lemmings.

Don Marsh and his team loaded onto LST (Landing Ship Tank) #1009 at Southampton docks late in the afternoon on D Plus 2, 8 June 1944.

Although the Americans weren't stationed at Bulford, they did occasionally venture across from Tidworth when dances were being held. This led to numerous fist-fights and other disturbances, normally over girls or regimental rivalry. Harry Ashford lived at Bulford Camp and he and his three sisters would also attend the dances, but his father would be on hand to escort his daughters home afterwards. Harry explains that the situation was worst when British and American paratroopers came into contact with one another. If a fight erupted the fuse would be pulled on the wall, but even in the dark the fight would continue. Men would punch at the light from a cigarette, on the assumption there was a face behind it. Harry recalls one fight starting after an American ranger boasted that his unit was 'the best'. He was taunted by a 'Brit' that the Americans couldn't 'even jump without two 'chutes', the American paratroopers having a back-up parachute that the British didn't, so if you jumped and the parachute failed to open you were dead. After a while, the Americans were banned from coming into Bulford Camp.

From the very start of the war, Everleigh Manor was requisitioned as a military hospital, but in April 1942 it was taken over by the Royal Army Medical Corps as a vaccine laboratory where important research and development was carried out. In the same year a temporary hospital was built close to the Manor for the United States Army, but at the end of the war this was demolished and the site restored to agricultural land.

Other American units based around the edges of Salisbury Plain, were close enough to use the downland facilities for training. These included the 4th Armored Division at Devizes, the 67th Tactical Reconnaissance Group of the Ninth United States Army Air Force at Middle Wallop, the 368th Fighter Group at Chilbolton, and the 366th and 370th Fighter Groups at Andover.

When the war was finally over, and the Americans had left the area, both Tidworth and Perham Down became transit camps for those women departing for the United States who had married American servicemen, many of whom had probably met their husbands at Red Cross dances at places like Tedworth House and the Cadre Club. From these camps, 640 GI brides would be processed, along with their 176 babies.

Chapter Five

Modern Times

At the end of the First World War, the military presence on Salisbury Plain was very quickly reduced, but at the close of the Second World War the situation was slightly different. Although many of the temporary wartime camps were dismantled, the British Army would retain much of its size. Britain now had military commitments all around the globe, and was about to enter the period known as the Cold War, so the country needed to maintain a large army of conscripted soldiers doing their 'National Service'. Other than those attending the various schools, the garrison troops on Salisbury Plain remained largely affiliated to the 3rd (UK) Division.

Post-War Expansion

Since the end of the war the garrison towns on Salisbury Plain have continued to change, with new barracks and married quarters constantly being built. There are now extensive civilian housing estates for people either working for the Ministry of Defence or in other forms of employment entirely. At Tidworth, military housing has included the Zouch and Mathew estates, as well as the Wylye Road development and housing in Plantation Road and Church Lane.

For soldiers and civilians alike, housing has improved greatly since the original Merthyr Tydfils were built. The garrison power station was closed in 1961, after which electricity was finally supplied from the National Grid. Even the oldest married quarters have been improved with modern facilities. In 1988, the Officers' Club at Tedworth House was closed, and the establishment was transformed into the garrison officers' mess.

At Perham Down, the Canberra estate was built in the 1960s, all the roads being named after Australian cities. Upnor Close and Wouldham Close followed in the 1970s. In 1975, the 22nd Engineer Regiment moved into Swinton Barracks, Perham, engineers replacing tank soldiers at a location not be confused with the barracks at Warminster that bore the same name. Both are named after the

87 *The Royal Inniskilling Fusiliers from Knook Camp, marching through Heytesbury for a church service at the parish church in 1958.*

'Father of the Royal Tank Corps', Major General Sir Ernest Swinton, who had also been an officer in the Royal Engineers. In this way, the link between the Royal Tank Regiment and Perham Down was maintained. The 22nd are still at Swinton today as part of the 3rd Division's 1st Mechanised Brigade.

Bulford's post-war building has included John French Avenue and three developments called the Australian Estate, the Irish Estate, and the Canadian Estate, the roads in each reflecting their names. A lot of this housing came into effect after a mains drainage scheme was built and connected to the Ratfyn treatment works at Amesbury in 1951. Beacon Barracks were rebuilt and re-designated Picton Barracks in the 1970s, undoubtedly named after Lieutenant General Sir Thomas Picton, who was killed at the Battle of Waterloo while in command of the 5th Division. But, as well as housing estates and barracks, Bulford has now acquired most of the amenities of any small rural town.

Larkhill perhaps saw the most significant development as, in the 1950s, the old timber and corrugated iron huts were eventually replaced with modern buildings. There was a complete redevelopment of the area north of the School of Artillery, above Brackenbury Road, which became Alanbrooke Barracks.

88 *Band of the 1st Battalion The Wiltshire Regiment at Le Marchant Barracks, Devizes, in November 1952. This was one of their last parades in battledress. Shortly after this photograph was taken the band was issued with infantry pattern blue number one dress. Arthur McKechnie is seen in the front rank with trombone lowered. The band master was L. Fereday LRAM.*

89 *The office of 22 (Gibraltar 1779-1783) Independent Locating Battery RA, at Roberts Lines, Larkhill. The battery was part of 3rd Division (Strategic Reserve). Its role in the 1960s was in counter-bombardment.*

Appropriately, Field Marshal Viscount Alanbrooke, KG, GCB, OM, GCVO, DSO, laid the foundation stone for the first building on 27 September 1960, the soldiers' dining hall.

To the east of Congreve Road Roberts Barracks was built, and during the 1960s there would be further improvements to the School of Artillery itself, including new headquarters and an instructional block. These new buildings were officially opened by General Sir Robert Mansergh, GCB, KBE, MC, the presiding Master Gunner, on 6 October 1966. Also built around this time was the Newcombe Hall, which has since been used as a conference centre, for talks and discussions, and to stage amateur dramatic productions by local groups. Today the garrison also includes Horne Barracks to the east of Knighton Down.

At Warminster married quarters were built at Elm Hill, Houldsworth Avenue, Wilson Square, Pepper Place and Goodwin Close, while houses for married officers were built at Oxendene and Elm Hill. During the 1950s, Boreham Hutted Camp was redeveloped with modern permanent buildings, and re-named Boreham Barracks in 1957 and, finally, Battlesbury Barracks in 1964. In 1978 Harman Lines was built on Sack Hill and named after General Sir Jack Harman, KCB, OBE, MC, who was Colonel Commandant of the Royal Armoured Corps at the time.

90 *Vehicles on the parade ground at Roberts Lines, Larkhill in the early 1960s.*

91 *The Packway Mess (Royal Artillery Officers' Mess) at Larkhill, photographed in the 1960s.*

The amenities of each garrison town also continued to improve, service clubs where soldiers could relax off-duty thriving in many locations. Today, garrison shops sport the Londis or Spar logo rather than that of the NAAFI. Over many decades, soldiers had become familiar with pubs on the Plain such as the *Bustard Inn* near Shrewton, or the *Saxon Warrior Inn* at Bulford. Today's favourite drinking haunts include the *Packhorse Pub* at Larkhill, *The Kiwi* at Bulford Barracks and *The Dog & Gunn* at Netheravon.

Many more churches have also been established or improved to cater for modern spiritual needs. At Tidworth, for example, the garrison church of St Michael began to hold monthly inter-denominational services in 1985 after the closure of St Andrew and St Mark in the same year, which had been run by the Church of Scotland, Methodist and Free Churches. The second garrison church of St Patrick and St George still administers to Roman Catholics within the garrison.

Larkhill's first permanent Catholic church, in Gore Road, dates from 1968 when it was dedicated to St Barbara and St Anthony by the Rt Rev. Gerard Tickle, Roman Catholic Bishop-in-Ordinary to the Forces, on 22 September. St Barbara is the patron saint of gunners everywhere. The Latin word for 'everywhere', 'Ubique', is the motto of the Corps, and is displayed on their badge between a crown and a cannon. At Bulford, the same Reverend Tickle also presided over the dedication ceremony of the present Catholic church of Our Lady Queen of Peace, on 12 January 1969. At Warminster, the first permanent Church of England garrison church was dedicated to St Giles on 9 June 1968 by the Venerable Archdeacon J.R. Youens, CB, OBE, MC, QHC. The foundation stone, which had originally come from St Giles' church at Imber, had been laid on 29 January

92 *Amalgamation parade attended by the Queen Mother at Tidworth, early 1959. The 3rd King's Own Hussars and the 7th Queen's Own Hussars were merged at this time to form the Queen's Own Hussars.*

1967 by Lieutenant General Sir Anthony Read, KCB, CBE, DSO, MC, who was commandant of the School of Infantry from 1959-62.

Delhi Barracks at Tidworth had become the garrison hospital in 1907, and it remained so until 31 March 1977, although a new hospital had been built on the same site in 1966. After the Americans left in 1945, and before it was turned into an officers' club, Tedworth House had been used to accommodate nursing staff from the hospital. Similarly, a building at the junction of Bazaar Road and Jamrud Road called Cavell House was also used as a nurses' home, but this closed in 1977, as well, and later became the garrison's sergeants' mess. The hospital building itself was subsequently used by several other units, including No. 2 Medical Group, Royal Army Medical Corps; Headquarters No. 1 Signal Group, Royal Corps of Signals; and Headquarters Army Fire Services, United Kingdom Land Forces.

In 1951 the Army Medical Stores at Ludgershall became the Army Medical Equipment Depot, and in 1969 changed its name again to the Defence Medical Equipment Depot. This depot provided all three services with medical equipment and supplies, whether they were at home or abroad on active service. It was responsible for purchasing, packing, storing, inspecting and maintaining all medical and dental equipment. It also trained medical storemen for the Royal Army Medical Corps, and pharmacy technicians from each of the services. Over the period from 1971 to 1982, the depot was rebuilt to comprise an extensive area of modern buildings that even had their own railway connection. Ludgershall has enjoyed extensive growth, with the building of many new housing estates, and in 1992 it incorporated Faberstown, which had previously been in Hampshire, the two now being regarded as the same community.

93 *John Chappell in the centre, with two of his crew: 2nd Troop, C Squadron, Queen's Own Hussars, based at Tidworth.*

However, by the 1960s, in common with the rest of rural Britain, the branch railway lines connecting garrison towns were being decommissioned. The railway station at Ludgershall closed to passengers in 1961 and completely in 1969. It was demolished in 1974 and houses were built on the site. Tidworth railway station closed in 1963, the tracks being removed in 1964. The site of the old station is now a supermarket. At Bulford, passenger trains were withdrawn in 1952, and goods traffic ceased in 1963. Work began on lifting the tracks in June 1965, and by September 1969 the new Salisbury Plain Area Works Office of the Property Services Agency had been built on the site of Bulford Village station. An old signal from the line was erected at the entrance to the approach road to these offices to commemorate their association with the railways. The only garrison town that now has a railway link is Warminster; for all the others, the nearest train stations can now be found at Andover or Salisbury.

94 *A lull during driving training on Salisbury Plain, with a Conqueror tank.*

95 *Conqueror tank of C Squadron, Queen's Own Hussars, exercising on Salisbury Plain in the late 1950s.*

One hundred years after the Army first began to build permanent barracks on their newly acquired land, the garrisons of Salisbury Plain are going through a period of major redevelopment called Project Allenby/Connaught. The MOD is working with Aspire Defence to provide 18,000 Army and civilian personnel with modern accommodation and working environments. These changes will effect Tidworth, Bulford, Perham Down, Ludgershall, Netheravon, Larkhill and Warminster. Many of the

96 *C Squadron Conqueror on the tank park at Tidworth.*

old barracks will be totally replaced and soldiers coming to the area will enjoy a better quality of life, both in their personal lives and while at work.

New Schools and Old

In terms of training, perhaps the greatest change in the area since the end of the Second World War is the formation of the School of Infantry. Although Warminster had grown as an armoured garrison in the 1930s, after the war it became a main training base for infantry. The School of Infantry moved there in 1945 from Barnard Castle in County Durham, where it had first been established in 1941 under the title of the Northern Command Battle School. Initially, the school was split over several locations, with Headquarters, Signals Wing and Tactical Wing at Warminster, the Small Arms Wing at Hythe in Kent, and the Support Weapons Wing at Netheravon. The Small Arms Wing transferred to Warminster in 1969. An administrative unit for officers and soldiers based at the school was called the Warminster Support Unit. This establishment catered for the staff and students of the Junior Division of the Staff College, formed at the school to train junior officers of all arms in tactics and fundamental staff duties.

A demonstration battalion was also attached to the School of Infantry, initially accommodated at Knook Camp until it was moved to Boreham Barracks, Warminster in 1964 (and later re-named Battlesbury Barracks). There was also an armoured squadron, detached from its parent regiment at Tidworth. These squadrons were located in Harman Lines, and artillery support was provided by a Field Regiment of the Royal Artillery from Larkhill.

The demonstration battalion was normally taken from a regular infantry regiment, although for a short period between 1978 and 1982 a composite battalion was experimented with. It was commanded by Lieutenant Colonel J.J. Caiger (Staffordshire Regiment) and, on disbandment, by Lieutenant

97 *HRH The Duke of Edinburgh visits his regiment, the Wiltshire Regiment (Duke of Edinburgh's) at Knook Camp, in November 1955. He is accompanied by Drum Major Pegg and, just in view, Lieutenant Colonel G. Woods, commanding officer of the 1st Wilts.*

Colonel P.J. Houghton (Welsh Guards). Ken Don served with this unique unit and explains:

> As a corporal in the Gordon Highlanders, I was posted to the advance party of the composite infantry demonstration battalion (IDB) at Battlesbury Barracks. The advance party actually formed up at Knook Camp until it took over duties from 1 DERR. In fact I was there from the beginning to the end of the IDB on extended service. There were probably no more than about 20 who did the whole tour, just over four years. In the battalion there were representatives of every battalion in the Army, plus Gurkhas. I was in C Company, which was mechanised. Whilst there, Imber village was virtually a second home. Imber Barn in fact was the main RV for feeding troops in the field with container meals. When the IDB was disbanded we were relieved by 1st Battalion Royal Welsh Fusiliers.

In 1991, the Combined Arms Training Centre (CATC) was established at Warminster, in accordance with the Options for Change Defence reorganisations taking place at the time. CATC took over the School of Infantry's All Arms Tactics Wing, and here soldiers would train in brigade and battle group formations, as well as armoured tactics and both armoured and infantry reconnaissance. CATC was also responsible for training soldiers to fight in built-up areas, a Built Up Areas Training Team utilising purpose-built constructions at Copehill Down.

The Battle Group Training Unit was based at Westdown Camp, Tilshead, and incorporated the Demonstration Battalion and Armoured Squadron of the old School of Infantry, which formed the basis of a permanent CATC Battle Group. To house much of this reorganisation, a new complex was built at Gale's Field. The group would include an armoured reconnaissance troop, an engineer close support troop and an REME light aid detachment, all working together.

In 1993, the headquarters of the Combined Arms Training Centre took over command at Warminster Garrison. The School of Infantry itself had become merely a part of this wider scheme, and was renamed the Infantry Training Centre (ITC). The barracks at Warminster became known as the Warminster Training Centre (WTC), and in 1995 the Infantry Support Wing at Netheravon closed and moved to the Warminster Training Centre.

Further developments included the building of the large Combined Arms Tactical Trainer between the Warminster Training Centre and Battlesbury Barracks. This is effectively a virtual battlefield for the 21st century. All types of weapons and tactics can be simulated in one of the most advanced military facilities in the world. The type of soldiering known as 'Synthetic Fighting' is considered crucial in the training for and development of future tactics and technologies. The unit can replicate any type of weapon system, from missile launchers and tanks to fighter aircraft.

To underline its importance, the Warminster Training Centre has, since April 2000, come under the command of a major general with the title Director General Training Support, a rank formerly held only by Directors of Infantry. In June 2000, CATC was renamed the Land Warfare Training Centre, and in 2002 it became the Land Warfare Centre, without question one of the most important military establishments in Britain today, encompassing much more than a centre at which to train infantrymen for the battlefields and streets of places like Iraq and Afghanistan.

98 *Field Marshal Sir John Harding talks to Band Sergeant E. Marsh of the Band of the Wiltshire Regiment at the School of Infantry, Warminster, in 1954. Also in the picture are Band Master L. Fereday and Lieutenant Colonel Woods.*

99 The *3 MK7 Mortar Locating Radar unit of 22 Locating Battery RA at Larkhill. Between October 1960 and February 1962, Grant Paton served as a subaltern with the battery and was in charge of a radar unit specialising in mortar location.*

One interesting feature of the Land Warfare Centre today is the Weapons Collection, which had formerly been located at the School of Musketry at Hythe. It moved to Warminster in 1969 and was officially opened by General Sir Anthony Read on 16 May 1970. It contains approximately 2,500 exhibits displayed in eight large rooms. Most of the weapons originate from after 1853, the year in which the collection was started, but there are numerous examples from before this time. In the main it covers small arms, mortars and anti-tank weapons from all over the world. The collection is open to the public, as well as members of the armed forces, but by prior arrangement only.

Although Warminster has become a main training centre for today's Army, Larkhill has retained its status as the home of the School of Artillery. As well as the new headquarters and instructional block, a Heavy Air Defence Compound was built in 1972 in which the Thunderbird 2 Guided Missile System was housed. When this was withdrawn in 1977, this enclosure near the junction between Brackenbury Road and The Packway was used as an anti-tank compound, and in 1977 the area was renamed Purvis Lines after Brigadier R.H. Purvis, CBE, the last Commandant of the School of Artillery, Manorbier.

100 The *3 MK7 Mortar Locating Radar and Tractor at Larkhill in the early 1960s.*

During the war years Larkhill became an obvious favourite with the royal family. Official visits since the war include 3 April 1973, when Her Majesty The Queen inspected guns, met staff and their families, lunched

101 *Royal Artillery firing the 205mm gun at Durrington Down in the early 1960s. It is believed this gun was capable of firing a nuclear shell.*

102 *The 205mm gun being towed onto the Gun Position on Durrington Down.*

in the Royal Artillery Mess and attended a meeting of the Royal Artillery Point-to-Point. To mark the 50th anniversary of the formation of the School, in 1970 the Queen conferred the prefix 'Royal', and the establishment became known as the Royal School of Artillery.

Today, the aim of the school is 'to provide the appropriate number of well trained and motivated individuals to meet the operational requirements of the Royal Artillery'. It is organised into a headquarters (which doubles as the Larkhill Garrison Headquarters), commanded by a full colonel, and five instructional branches. Since 1994, the school has also been the location of the Headquarters of the Director Royal Artillery. The five instructional branches cater for Gunnery, Air Defence, Command and Information, Systems, and Tactics and Young Officers. To aid the training programme, the school is supported by the 14th Regiment Royal Artillery. Between them, the five instructional branches and the training regiment run more than 200 courses a year, ranging from 'Gunner Basic' to 'Commanding Officer Designate'. Much of the training today is achieved through high-tech simulators and state-of-the-art computer programmes.

Work, Rest and Play

Since the end of the war, tens of thousands of military personnel have lived and trained at the garrisons on Salisbury Plain. Some were regular soldiers, others were doing National Service or were members of the TA. Some remember the Plain with affection, others with dread.

Mr G. Gilbert, for instance, did his National Service in the Somerset Light Infantry. He was based at Bulford Camp, and was taken from there to Salisbury railway station on 17 September 1952, from where a train took him to the dockside in Liverpool to board the troopship *Empire Pride*, bound for Singapore and service in Malaya. After completing National Service you were expected to do a further three years in the Territorial Army, and during this time Mr Gilbert returned to both Bulford, for some of his training, and Tilshead to use the rifle ranges. On one occasion he took part in a day and night exercise intended to simulate the actions to be taken after an enemy attack using atomic bombs. He explains:

103 *Farewell parade at Candahar Barracks Tidworth, as C Squadron of the Queen's Own Hussars prepare to leave for Aden in February 1960. John Chappell is the tall figure in the front row nearest the camera.*

> We were dropped off early evening before dark, to dig in and then cover ourselves with galvanise sheets, which were supposed to act as a shield against radiation. It was September time. Have you ever tried digging a trench on Salisbury Plain? We dug down about six inches and then hit solid rock. So we lay there all night with the galvanise sheets on top of us. At first light, smoke screens were dropped, as the sound of tanks was getting ever nearer. 'It's time to get out,' we all thought, and anyway, we had been 'so called' beaten by the enemy. So we got onto lorries to be taken back to the NAAFI. We never used our rifles, we just made banging noises instead, and kept our rifles clean for inspections back at camp. Pull through down the spout and a duster around the workings. What a waste of time and money!

David Piper was born in 1946 in the village of Shrewton, four miles from Larkhill, and four miles from the Tilshead camps. His father had been in the Royal Artillery at Larkhill during the 1930s. At the age of 12 he joined the local Army Cadet Force platoon, which was attached to the 4th Battalion The Wiltshire Regiment (TA). Cadets from all over Wiltshire, and the wider country, have often enjoyed the unique training facilities that Salisbury Plain has to offer, as Mr Piper describes:

> I spent most of the next six years exercising on Salisbury Plain and rose to the exalted rank of Sergeant. We had Area weekend camps at Tilshead – 'C' Camp and Westdown. Half of that time was spent in Shrewton attached to 4th Wilts., the other half at Larkhill attached to the Royal Artillery. During the 1950s, our unit was invited to witness the firing of the Honest John missile at Larkhill. We gathered one afternoon with all kinds of NATO officers, military officials, Artillery brass, and Sandhurst Cadets, on the downs behind the School of Artillery. It took a long time to set up the rocket, but when it was fired it made a tremendous explosion and shot off across the countryside to the impact area near Warminster in seconds. On the same afternoon we were

104 *A rolled tank at Tidworth. This was a Mark 5 Centurion tank of the Queen's Own Hussars and the incident occurred shortly after their arrival at Tidworth. It happened on the tank park of Candahar Barracks while unloading from a tank transporter. The trailer the tank was on tipped to one side allowing it to topple over.*

> treated to a battalion assault by the 1st Battalion, Parachute Regiment, on a Gurkha unit acting as the enemy. The paratroops were dropped right overhead and we saw them land, form up, and move into action with a commentary all the way.
>
> During the summer evenings of the 1950s and 1960s, we would sometimes see a number of Beverley aircraft flying over. This was usually an indication that there was to be a training parachute drop on the downland. The usual procedure was for six or seven Beverleys to fly over the downs, circle round, and then make their dropping run. We would rush for vantage-points to see the spectacle of 600 paratroopers and their vehicles being dispatched on the downs. During the summer months TA units, particularly from the North of England and Scotland, held their annual camps at Westdown in Tilshead, and those that brought their bands with them used to parade them at the Catherine Wheel Square in Shrewton. I distinctly remember the Argyll and Sutherland Highlanders bringing their pipe band to the village. Artillery gunfire always played a backdrop to the villages around the area. When travelling from Shrewton towards Larkhill the red flags were often flying, and sometimes we could see the guns out on the downs firing across towards Imber.

And it wasn't merely soldiers and cadets that enjoyed these facilities. John Fellows is a retired Wiltshire police officer, who worked with and trained police dogs while based at Devizes. He ended his career as the force's instructor, and here explains how, when he was still only a constable, his entire dog section nearly got wiped out on Imber ranges:

> Our Dog Section used to use Salisbury Plain for training purposes and I can recall going to Imber village to assist the Army with riot/tactical training. My police dog has bitten many a squaddie on their legs and backsides! The incident I refer to took place in approximately 1984. We had permission from a soldier on guard duty to use an area of the Plain to do some 'team

105 *Honest John long-range artillery missile being tested by the Royal Artillery at Larkhill in 1961.*

106 *Honest John missile being fired at Larkhill. Developed in the USA, it was capable of carrying an atomic or a high explosive warhead.*

107 *The Royal Artillery practise night firing with the Honest John missile at Larkhill in 1961.*

tracking'. This is where one of us goes off and an hour later we have to track him with the dogs following his scent. There would have been about six of us with dogs. As we started off after our quarry we heard the sound of explosions, nothing surprising on the Plain! We had progressed about two miles and these explosions were getting nearer and nearer. Suddenly we all dived for cover as the shells were landing near us. While we were sheltering an Army Landrover pulled up with an officer shouting, 'What the **** are you lot doing here, you could all have been killed?' He asked who was in charge and we all pointed at our sergeant, who was taken away in the Landrover.

The sentry who let us onto the Plain was apparently court-martialled. Our sergeant got a right bollocking, and we forgot all about our quarry, who by now was hiding in a small copse on the Plain waiting for us to find him. We returned to our headquarters at Devizes, where some three hours later the quarry appeared, really cheesed off! Although we all saw the funny side of the incident, we also realised that half of the Wiltshire Police Dog Section had nearly got blown away.

John Grosvenor has a happier and quite unusual memory of his time on Salisbury Plain, aged 19 and serving with the 1st Battalion The Royal Scots (The Royal Regiment). His regiment was stationed at Assaye Barracks, Tidworth in 1965 when it was called upon to supply extras for the film *Help*, starring The Beatles, which was partly filmed in the area:

108 *The Beatles on Salisbury Plain, during the making of the film* Help *in 1965. John Grosvenor recalls how his battalion of the Royal Scots were employed as extras in the film. The band performed behind a screen, which protected them from the wind.*

> We were driven onto the Plain in three-ton trucks from Tidworth. The Beatles would be driven from Salisbury, I believe in a Rolls Royce, to the treeless location for filming. Other Royal Scots and I were to lie between Centurion tanks in a circle around The Beatles, who played a couple of hit songs behind a glass screen to keep the wind away from them. The tanks were from 3rd Royal Tank Regiment. We got on great with Ringo, and Colour Sergeant Davy Greenhill lent him his Tam-o'-shanter to wear during the filming. In the film Ringo wears this with the cap badge at the front of the head and not above the left ear as should be worn. The badge is the Lowland Brigade badge, as Scottish regiments were split into Lowland and Highland brigades. The Royal Scots re-badged back to their original cap badge in 1969.

In the 1960s, it was decided to revive the Tidworth Tattoo, by attaching an afternoon event to the end of the Army's annual three-day horse trials. The trials had first taken place at Tedworth Park in 1954 as a one-day event, becoming a three-day event in 1956. The first of the tattoos took place on Sunday 22 May 1966, but unfortunately very high winds caused havoc. The massed bands could hardly be heard, and the programme highlight of a free fall parachute display by the Red Devils had to be cancelled, as the aircraft were unable to take off from Netheravon. Undaunted, further tattoos took place over the next ten years, with the very last one being held on 31 May 1976.

The horse trials themselves were also discontinued between 1978 and 1981, when they were re-established thanks to the efforts of Major R.D. Baker of the 16th/5th The Queen's Royal Lancers, who decided to reinstate them as a one-day event after he was appointed to a position within the Headquarters South West District in 1979. These trials, still held at Tedworth Park today, proved very popular and are now recognized as one of the best one-day trials held anywhere in the United Kingdom. Similarly, point-to-point fixtures at Larkhill, on Knighton Down, are also well supported by families of service personnel and the general public. The first ever point-to-point at Larkhill was held on 23 April 1947, and really came to recognition after the visit of Her Majesty The

109 *Remains of a tank pictured on Salisbury Plain in the 1950s, used at the time for tank gunnery target practice.*

Queen and the Duke of Edinburgh on 21 February 1953, to watch the United Services Point-to-Point.

Tedworth Park also remains the home of the Tidworth Polo Club, which accommodates numerous competitions throughout the year between the Army and Navy, as well as the armies of foreign countries. The club also entertains teams from schools, such as Millfield and Taunton. The military cups and trophies that are up for grabs include the Queen's Royal Hussars Cup, the Queen's Royal Irish Hussars Cup, the King's Royal Hussars Cup, the Light Infantry Cup, the Cavalry Cup, The Royal Artillery Harriers Cup, the Indian Cavalry Officers' Trophy, the Captains and Subalterns Trophy, and the Queen's Royal Lancers Cup.

110 *Corporal Rocket at Larkhill in the early 1960s. This was the first guided missile capable of carrying a nuclear warhead. It had a range of 75 nautical miles and was developed by the Americans as a tactical nuclear missile for use in the event of Cold War hostilities.*

The Royal Artillery has had its own hunt in one form or another since 1907, and in 1948 Colonel W.C.P. Richardson, DSO, the Commandant of the School of Infantry's Tactical Wing, was instrumental in the formation of the School of Infantry Beagles, aided by Major D.B. Rooke, MC, an instructor at the school who became the Hunt's first Master. It hunted largely on the MOD ranges between Warminster and Netheravon, although it has been known to hunt as far west as

111 *A Corporal Rocket on its transporter at Larkhill.*

Wincanton and as far north as Lacock. By 1985, the Hunt had become diluted, in military terms, by the inclusion of many civilian members, both of the hunt and on the committee, and so changed its name to the Wiltshire and Infantry Beagles.

112 *Mess halls at New Zealand Farm in the 1950s.*

In 1951 the School of Infantry Saddle Club was founded by the Commandant, Brigadier R.E. Goodwin, DSO, with the assistance of Lieutenant Colonel J. Mogg, DSO, and Major J.B. Dye, MC. The name was later changed to the Infantry Saddle Club, and at the start was set up exclusively for Army personnel wishing to learn how to ride. It still exists today just off the Imber road, but is simply called the Warminster Saddle Club and is now open to both civilians and military personnel. The club has indoor and outdoor riding schools, as well as show jumps and a cross-country course. Events are run at the school throughout the year, and include show jumping, hunter trials, dressage, the annual summer show and various British Horse Society events.

113 *Interior of accommodation hut number 2 at New Zealand Farm in the 1950s. John Webb lived both here and in hut number 3 (of which he was in charge) while serving at the camp.*

114 *An RAF policeman controls the traffic at the entrance to New Zealand Farm.*

If Salisbury Plain was to lose one of its major annual events with the demise of the Tidworth Tattoo, this was compensated for somewhat with the first ever 'Larkhill Day'. This was held on 27 and 28 July 1962, and the public was entertained with static displays of guns and equipment and an impressive demonstration of artillery fire on the ranges. This was to become a regular event, held every four years, and in 1970 its title was changed to 'Artillery Day'. The last Artillery Day at Larkhill was held between Thursday 29 June and Saturday 1 July 2000, the event in 2004 being cancelled due to the deployment of troops in Iraq. Arena displays included the Flying Gunners Motor Cycle Display Team, the Royal Artillery Freefall Parachute Display Team, the Royal Artillery Band, the Musical Drive by the King's Troop Royal Horse Artillery, the Army Tug of War, and a parade of the Royal Artillery Foxhounds. The finale was a fantastic display in the live firing arena which showed the different ways in which guns and rocket launchers could deploy to the battlefield and engage the enemy. It included Hercules transport aircraft from RAF Lyneham landing and taking off again, and a demonstration of the Army's latest Apache attack helicopters.

115 *New Zealand Farm football team, with John Webb in the front row far left. The team would play other sections of the RAF from Boscombe Down, and teams from the Royal Navy.*

Apart from grand events like 'Artillery Day', and public meetings such as the point-to-point, the garrison communities on Salisbury Plain offer the soldiers living there a range of different sporting activities. There are rugby, football and cricket tournaments, while at the other end of the sporting spectrum there is the Army Parachute Association at Netheravon and the Service's Dry Fly Fishing Association, also at Netheravon, which looks after six and a quarter miles of river and numerous artificial lakes in which it rears its own fish.

Changing Hands

After the Second World War, with the exception of Boscombe Down perhaps, the importance of the airfields on Salisbury Plain began to wane. They were also about to change hands, some going from the control of the Royal Air Force to that of the Army.

RAF Upavon continued as a flying school until 1946, when it became the headquarters of No. 38 Group. However, when it also became the headquarters of Transport Command in 1951, No. 38 Group was disbanded, and even though it was reformed again in 1960 it was moved to Odiham, as there was not enough space at its former home for both establishments to co-exist. This allowed the facilities at Upavon itself to be enlarged and improved. Transport Command in August 1967 became Air Support Command.

On 16 June 1962, the airfield celebrated the 50th anniversary of the establishment of the Central Flying School, with a flying display and static exhibits which were visited by the Duke of Edinburgh. The highlight of the occasion was a fly-past by No. 1 Squadron RAF, for which the salute was taken by Marshal of the Royal Air Force Sir John Salmond, who had attended the very first course at the Central Flying School in 1912.

Further changes within the RAF would continue to affect Upavon. For instance, on September 1972 Air Support Command was amalgamated with Strike Command, and Upavon became the headquarters of No. 46 Group. Then, on 10 November 1975, No. 46 Group was itself amalgamated into No. 38 Group. Thus, 38 Group returned home

116 *'Working at New Zealand Farm we had extra cold weather clothing,' explains John Webb, 'including Wellington boots, sea boot stockings, string vests, polo neck jumpers, gauntlets, leather body warmers, waterproof anoraks with hoods and scarves, as modelled here by Dave Young.'*

to Upavon, and was for a while the largest single group in the Royal Air Force before, in November 1983, it was finally absorbed into No. 1 Group.

Over the years, the RAF had been a major employer of local people at Upavon, had provided custom to local businesses and had filled the village school. But in 1993 the RAF handed over the station to the Army, who renamed it Trenchard Lines. Since then the airfield has been used less and less for flying, as the site has been converted mainly to an administrative centre. Based here are the Office for Standards of Casework (Army), the HQ of the Adjutant General, Service Children's Education (UK), the Army Training and Recruiting Agency, and the HQ of the Provost Marshal's Office. The airfield also houses RAF No. 622 (Volunteer) Gliding School and the Army Gliding Association. Occasionally, Hercules aircraft from RAF Lyneham and the odd Chinook helicopter from RAF Odiham might still be seen landing on the airfield.

RAF Netheravon closed as an airfield in 1950 and from 1 December 1950 to 1962 it was an RAF Police Training Centre, housing both the Police Dog Training School and Police Driving School. In 1962 the Lower Airfield Camp became an Army transit area for units exercising on Salisbury Plain, while the Upper Camp fell into disrepair until it was partly taken over in 1963 by the Army Free Fall Parachute Association, which later became the Joint Services Parachute Centre.

In 1964 Netheravon became the Headquarters of No. 651 Squadron of the Army Air Corps, which was of course descended from the very first Air Observation Post squadron formed at Old Sarum back in 1941 under the command of Lieutenant Colonel Henry Charles Bazeley. The Army Air Corps Centre was formed at Middle Wallop in Hampshire on 1 September 1957, and the School of Army Aviation was set up in August 1965. Netheravon's major station unit today is No. 7 Regiment AAC (V), which was formed in 1969. This is a Territorial Army regiment affiliated to the Army Air Corps, whose main role is to provide helicopter liaison flying for the Regular Army and general aviation support for the Territorial Army. Above the main entrance to the Officers' Mess at Netheravon is a large carved set of Royal Flying Corps wings, which go at least some way to indicating the station's long and unique association with Army flying.

117 *Demob day for John Webb, 24 September 1956, outside hut 2, New Zealand Farm.*

118 *Thunderbird Rocket at Larkhill in 1961.*

Although Old Sarum also changed hands in 1971, and became the base of No. 622 Gliding School until October 1976, it finally closed its doors to military use in 1979. So, of all the airfields on Salisbury Plain, this left only Boscombe Down under the control of the Royal Air Force.

Anyone driving west along the A303 today and descending the hill towards the Countess roundabout must be aware of the huge hangar at Boscombe Down to their left. The massive structure is a Weighbridge Hangar, built in 1954 to permit the largest aircraft to be worked on under cover as well as provide accurate weighing facilities. In the 1960s and early 1970s, Boscombe Down grew to such an extent that it was organised into eight divisions: Armament, Engineering, Flying,

119 *Hardened OP on Salisbury Plain in the 1960s, possibly on Honeydown Ridge.*

Navigation and Radio, Performance, Photographic, Technical Services and Trials Management. Today, RAF Boscombe Down is mainly involved with testing and evaluating new aircraft and aircraft equipment.

Back in the 1950s and 1960s, it supported the out-station at New Zealand Farm, and it was here that John Webb served much of his four years as a regular in the Royal Air Force. John first went to Cardington in Bedfordshire for kitting out, and from there was posted to Bridgnorth in Shropshire for square bashing. After passing out, he went to Weeton in Lancashire to do a mechanics course lasting 24 weeks, before finally arriving at Boscombe Down. However, his first period of service in Wiltshire was short-lived:

> After about ten months in the MTR&S, they decided it should be manned by civvies and all the RAF boys were sent to Weeton again for driving tests. This second stint at Weeton lasted about 10 days. We had three tests, one in a Hillman Minx, one in a 15 cwt Bedford and the third in a three-ton Austin Commercial. All of these I passed and was sent back to Boscombe Down. Six months later I was posted to New Zealand Farm. This being about ten miles from my home, I was quite happy to serve out my time there as a driver and mechanic. We were an outpost of Boscombe Down and took photographers and other personnel over the Imber Ranges to collect parachutes that had been dropped with cameras attached to them, to check their deployment. We also had night ops. Being an outpost of Boscombe Down, we were required to take RAF personnel to the barriers on Imber Ranges to relieve the civvies. Then, every two hours, we went round with tea and sandwiches and their reliefs. After the ops we picked them all up again. This usually lasted most of the night.
>
> Whilst at New Zealand Farm we had a good scheme going, catching rabbits on the ranges and taking them to Devizes market. That was until myxomatosis came along. We all clubbed together and had a gold and silver wire badge made by a firm in Devizes to put on our blazers, which had a rabbit sitting on a bomb, with a patch over one eye.

120 *An Army Lynx helicopter from Middle Wallop photographed at Deptford Down from the public byway on 29 November 2006.*

John Webb was demobbed at New Zealand Farm on 24 September 1956. Today, the same site is an Army camp and John states that it featured in the first series of 'Bad Lads Army' on television.

Some of the main units at Boscombe Down today, Salisbury Plain's remaining RAF outpost, include the Tornado In-Service Software Maintenance Unit, the Harrier Software Maintenance Unit, the Directorate of Flying, the Joint Test and Evaluation Group, and the Empire Test Pilot School.

Mock Villages

After the village of Imber was requisitioned by the War Office in 1943, it was used for a time by the American Army to train in street fighting, in order to prepare soldiers for the type of resistance they might encounter in the towns and villages of France. After the war, the authorities retained the village, and although a number of people have fought very hard over the years to return it to its former inhabitants, it has remained in the hands of the MOD ever since. One of Imber's champions was councillor Austin Underwood. He organised a number of protest rallies in the 1960s and, the village is now open to the public

121 *EH101 Merlin photographed from the public byway at Deptford Down on 10 January 2007.*

on certain occasions during the year. These include an annual service on the Saturday nearest St Giles Day, 1 September.

The village itself has become almost unrecognisable, as properties fall into complete disrepair. The British Army continued to use Imber, to train its own soldiers in urban warfare, and in the 1970s the crumbling cottages were overshadowed by a series of new buildings, that were purpose-built for fighting in built-up areas. These blocks of imitation houses, which became known as 'Chester Flats', were used to prepare soldiers to fight in Northern Ireland, a new wave of terrorist activity meaning that a large military presence was required in the province. The imitation houses displayed the trappings of an Ulster town and, among the dozen or so blocks, was a police station, job centre, the *Angel Inn* and Big Jim's Café.

Similarly, at Copehill Down near Chitterne, a brand new village built in 1986 was made to look like a Bavarian town, as at this time the Army was preparing to fight the Soviet Union in the streets of Eastern Europe. The training village is known as a FIBUA (Fighting In Built-Up Areas). Indeed, it is the principal FIBUA in the United Kingdom and is part of the Urban Operations Wing run by the Land Warfare Centre at Warminster. The town is built on the lines of six streets and contains somewhere in the region of 80-plus buildings. Today, it includes an Iraqi street scenario.

Although the mock village didn't affect any local housing, Sue Robinson explains in *Chitterne – a Wiltshire Village* that the local people were initially opposed to its construction:

> The training village, or FIBUA, was built in 1986 on Copehill Down against the wishes of the majority of the people of Chitterne. There were protests and meetings and villagers appeared on the TV news, but all to no avail. The villagers were told that a training village was needed to prepare troops for the possibility of a third world war that was expected to take place on the continent in Saxony. Thus the FIBUA was to be built to resemble a village of that region, high on the open plain, at a rumoured cost of £10,000,000.
>
> Despite the protests the planned building work went ahead and concrete houses soon sprouted up, where before there had been just the wild chalkland flora. Swathes of trees were planted in carefully planned groups in the surrounding area and rights of way were diverted around the new development. The villagers were worried that squatters or travellers might move into the empty buildings huddling around the taller replica church. But reassurances were given that there would be a continuous presence guarding the site, and slowly we became used to the FIBUA village and no longer stared at it as we drove past.

Only three years later the Berlin Wall came down and the purported scenario for hostilities disappeared. The temptation for Chitterne people to say 'we told you so' was very strong, and many thought the MOD had created a white elephant. But other threats have emerged, so the FIBUA village has been far from redundant. It has been adapted as necessary for training troops for fighting in Bosnia and, lately, Iraq.

Uniting the FIBUA at Copehill Down and the Combined Arms Tactical Trainer at Warminster, the Land Warfare Centre is well placed to prepare the modern British Army to deploy to urban conflicts around the globe. And villages like Chitterne rarely see a tank or armoured vehicle in its streets today, as the MOD constructed its own road, the Southern Range Road, to link all the camps on this part of the Plain. The road helps preserve the natural life and history of the Plain by keeping Army traffic to specific routes.

As for the village of Imber, the only original building which remains intact is St Giles' Church and, although it was made redundant in 2002, this grade one listed building is now vested into the guardianship of the Churches Conservation Trust. It is presently undergoing a three-year restoration programme to be completed by Easter 2008.

Salisbury Plain in the 21st Century

Since the Army first began to garrison Salisbury Plain, many different regiments have come and gone. In 2008, other than those based at the Land Warfare Centre at Warminster or the School of Artillery at Larkhill, the Plain's major units include the following:

- 1 Royal Horse Artillery, at Assaye Barracks Tidworth
- 2nd Royal Tank Regiment, at Aliwal Barracks Tidworth
- 1st Battalion Princess of Wales's Royal Regiment, at Bhurtpore Barracks Tidworth
- 1st Battalion Staffordshire Regiment, at Mooltan Barracks Tidworth
- 6 Battalion Royal Electrical and Mechanical Engineers, at Jellalabad Barracks Tidworth
- B (16) CS Medical Squadron, at Candahar Barracks Tidworth
- 1 Mechanised Brigade HQ and Signal Squadron (215), at Jellalabad Barracks Tidworth
- 22 Engineer Regiment, at Swinton Barracks Perham Down
- 26 Engineer Regiment, at Corunna Barracks Ludgershall
- 3 (UK) Division HQ and Signal Regiment, at Picton Barracks Bulford
- 12 Mechanised Brigade HQ and Signal Squadron (228), at Ward Barracks Bulford
- 1st Battalion 22nd (Cheshire) Regiment, at Kiwi Barracks Bulford
- 19 Tank Transporter Squadron, at Ward Barracks Bulford
- 158 Provost Company Royal Military Police, at Gordon Barracks Bulford
- 4th Battalion The Rifles, at Bulford Camp

The Ministry of Defence and the armed forces take their role as custodians of this historic landscape very seriously. Although the Army Training Estate is controlled from its headquarters at the Land Warfare Centre in Warminster, since 1995, the Salisbury Plain Training Area itself has been managed from the joint Military and Defence Estates Headquarters at Westdown Camp Tilshead. In view of the destructive power of the weapons being used in the area, it might seem odd to state that the Ministry of Defence is committed to conservation. But, in fact, the ancient monuments found on the training area are all protected, as are many habitats that support the area's indigenous flora and fauna. Indeed, it would be fair to say that the presence of the Army on Salisbury Plain has helped protect wildlife. Animals and plants exist here that in other cultivated areas have all but disappeared.

Salisbury Plain is a Site of Special Scientific Interest and a haven for many of Britain's most endangered species, such as the Stone Curlew, the Marsh Fritillary

122 *Derelict remains of a Chieftain main battle tank on Salisbury Plain near Imber, intended for target practice.*

Butterfly, and many rare orchids and arable weeds. The Ministry of Defence works in partnership with bodies such as English Heritage, Natural England, the Royal Society for the Protection of Birds, and the Wiltshire Wildlife Trust. The Great Bustard, for instance, became extinct in Britain in the 1840s, but in recent years has been re-introduced to Salisbury Plain. Although it's still early days, the experiment has met with some success. Hopefully, in years to come the Great Bustard pictured on the sign of *The Bustard Inn* won't be the only example of this magnificent bird observed in the area by soldiers and civilians. In the future, as in the past, many thousands of soldiers will train on Salisbury Plain, but soldiers today are taught to understand the value of preserving the estate for future generations.

123 *This unusual photograph might look like a waterfall, but in fact it shows tank tracks in the chalk at Clarendon Hill.*

Index

Illustrations are indicated in **bold** type